Mastering 37 WhatsApp Tricks

Mastering 37 WhatsApp Tricks

Zico Pratama Putra

Kanzul Ilmi Press
2017

Copyright © 2017 by Zico Pratama Putra

All rights reserved. This book or any portion thereof may not be reproduced or used in any manner whatsoever without the express written permission of the publisher except for the use of brief quotations in a book review or scholarly journal.

First Printing: 2017

ISBN-13: 978-1544876214

ISBN-10: 1544876211

Kanzul Ilmi Press
Woodside Ave.
London, UK

Bookstores and wholesalers: Please contact Kanzul Ilmi Press email

zico.pratama@gmail.com.

Trademark Acknowledgments

All terms mentioned in this book that are known to be trademarks or service marks have been appropriately capitalized. WhatsApp, Inc., cannot attest to the accuracy of this information. Use of a term in this book should not be regarded as affecting the validity of any trademark or service mark.

WhatsApp is registered trademark of WhatsApp, Inc.

Unless otherwise indicated herein, any third-party trademarks that may appear in this work are the property of their respective owners and any references to third party trademark, logos or other trade dress are for demonstrative or descriptive purposes only

Ordering Information: Special discounts are available on quantity purchases by corporations, associations, educators, and others. For details, contact the publisher at the above-listed address.

Dedication

Mila –

Because of her continued support & love throughout the writing of the book and also within my own life, she's helping me in more ways than anyone else.

Contents

Introduction 1

37 Tips and Features in WhatsApp Not Many People Know 3

   1. Create Status in WhatsApp 4

   2. Safer Feature Two-Step Verification 5

   3. Add Robot 7

   4. Change the WhatsApp scene 10

   5. Make WhatsApp Account Without a Phone Number ... 16

   6. Backup WhatsApp Conversations into TXT Format 20

   7. Save Internet Quota While Using WhatsApp 25

   8. Turn off the mark "Read" Blue Tick 31

   9. Hiding Status "Last Seen" 32

   10. Knowing Your Message has been Read While Blue Tick is being Turned Off 33

   11. Edit photos ala Snapchat Before Send 36

   12. Give Star Signs 38

   13. Mention Friends in Group 39

   14. Replying to a Message Drowned in Group 40

   15. Turn off Automatic Download 42

   16. Blocking Contacts 43

   17. Video Call 44

   18. Now You Can Send Files Word, Excel and PowerPoint via WhatsApp 46

   19. Automatic Encrypted Messages WhatsApp 48

20. Quick Reply Notifications 49
21. Can Send Text Bold and Italic 50
22. Retrieve Deleted Messages 52
23. Knowing Someone Location Through WhatsApp 57
24. Read Messages WhatsApp Without Eliminating Blue Tick 61
25. WhatsApp enables users to check the message read time 68
26. Using 2 in 1 Smartphone WhatsApp Account 70
27. Tie WhatsApp to your mobile number 80
28. Make Your WhatsApp Friend Damage (Crash) and Unused 81
29. Protect your privacy by disabling Preview 82
30. Use WhatsApp on Computer or Laptop 83
31. Sending & Scheduling WhatsApp Messages Automatically 87
32. How to know if I'm blocked on WhatsApp 95
33. Mute group chats if you get bored 98
34. Backup WhatsApp message to Cloud Storage 100
35. Use WhatsApp without Mobile Number 107
36. Maximize WhatsApp New Features 110
37. Five Fraud Type Strike WhatsApp Users 114
About the Author 119
Can I Ask a Favour? 120
Index 121

INTRODUCTION

WhatsApp is a messenger that aims to replicate the SMS experience, but avoid the expense of SMS. Part of what makes SMS popular is that it rides on top of the phone numbers that you can also use to make a phone call. So, you can exchange a phone number with someone, and then you have a choice of reaching that person via a telephone call, or an SMS text message. WhatsApp expands on that idea, so that you can now also reach a person via WhatsApp if you have their phone number. This phone number matching done by WhatsApp is not some optional feature -- it's a core aspect of using WhatsApp, as there is no other way of finding people than having their phone number.

The book is deliberately structured to expand your idea in using the WhatsApp features. You can take what you are looking for and apply it accordingly. WhatsApp and most other technologies are best learned by doing, so don't just read this book and do nothing.

This book is published with the title "Mastering 37 WhatsApp Tricks". On this issue, we have added some of the latest features of WhatsApp with backup conversation, video call, additional 3rd party software, etc. as well as make improvements layout and cover design to be neater and interesting to read.

Hopefully, with the existence of this book, you have a lot of ideas to maximize your communication experience through WhatsApp.

Write to be understood, speak to be heard, read to grow.
Good luck!

Mastering 37 WhatsApp Tricks

37 TIPS AND FEATURES IN WHATSAPP NOT MANY PEOPLE KNOW

WhatsApp is very brilliant in presenting a variety of new features to serve users. Unfortunately, there are still many users who do not know or do not use it.

In fact, you can use it and make it easier in many ways. Among them are to improve *chat* experience, maintain privacy, reduce the consumption of data, and much more.

Therefore, we have been updated the following 37 WhatsApp features and tips that surely none you already know yet or have not been used.

1. CREATE STATUS IN WHATSAPP

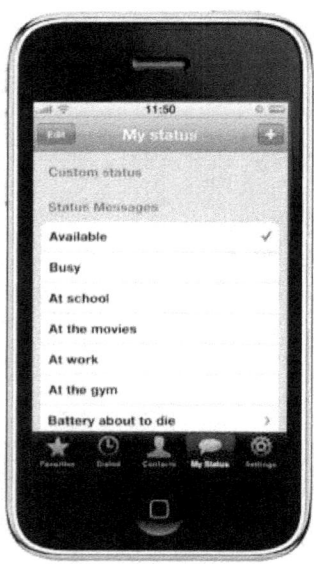 

WhatsApp again launched an interesting feature that can be used by many people. Almost like Snapchat and **Instagram Stories,** you can change the status to be more attractive than ever.

Not only to share your writing, you can also use this feature to share pictures or videos with your friends.

2. SAFER FEATURE TWO-STEP VERIFICATION

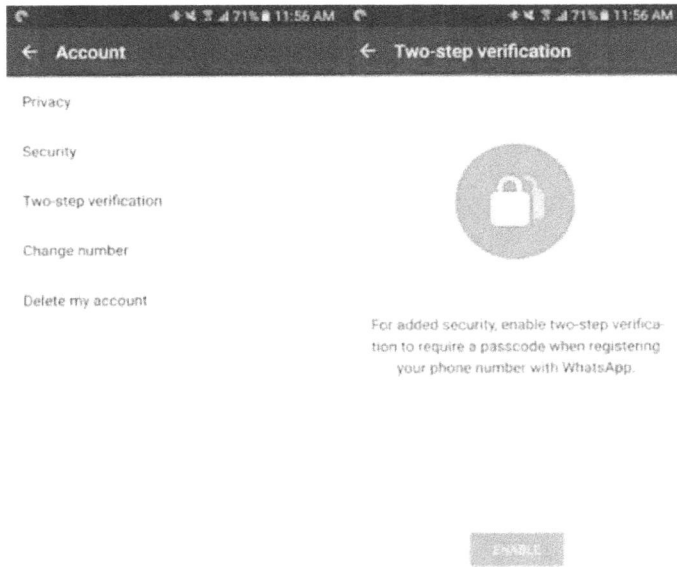

WhatsApp has never stopped for innovation in improving user's experience and safety. Do you know, the *two-step verification* feature is already implemented in the WhatsApp application, both for Android and iOS.

Two-step verification is a feature that requires you to enter a special password/passcode to be able to access information from your account. For you who care about privacy, set it now, here's how:

- Go to the menu **Settings> Accounts> Two-step verification> Enable.**

- There you should set a 6-digit *passcode* you want to use.
- After enabling the feature, you will be prompted to enter *a passcode* every seven days.

3. ADD ROBOT

Are there any of you are using Telegram? Compared with WhatsApp, Telegram has an advantage since it has *Bot* features that can be used to automatically execute a command.

In fact, *Bot* Telegram can be invited to play a game. Well, to make WhatsApp more sophisticated and exciting, you can also add **Bot** on **WhatsApp,** here's how!

- **Download the qeuBot application from Google Play**, and install as usual.

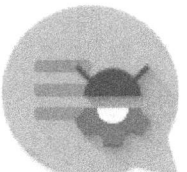

- **QeuBot** is a useful application to add Bot in WhatsApp. You can use the bot to search for information on IMDB movies, weather, latest news, GIF images, or even performing mathematical calculations.

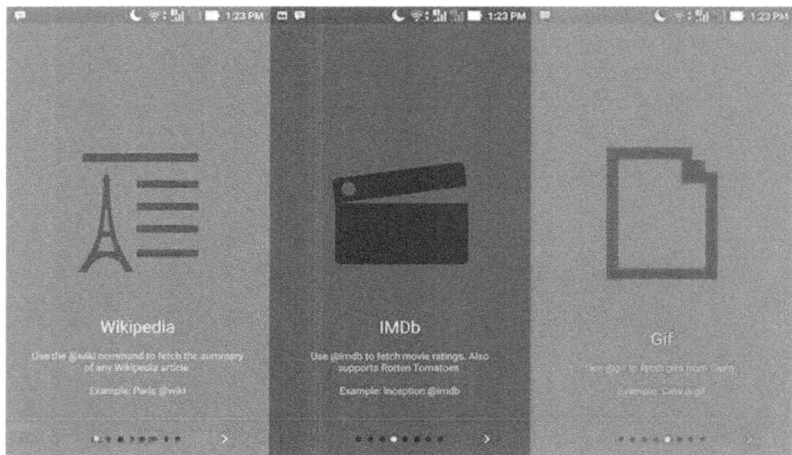

- After qeuBot has installed, activate the **Accessibility Service**.

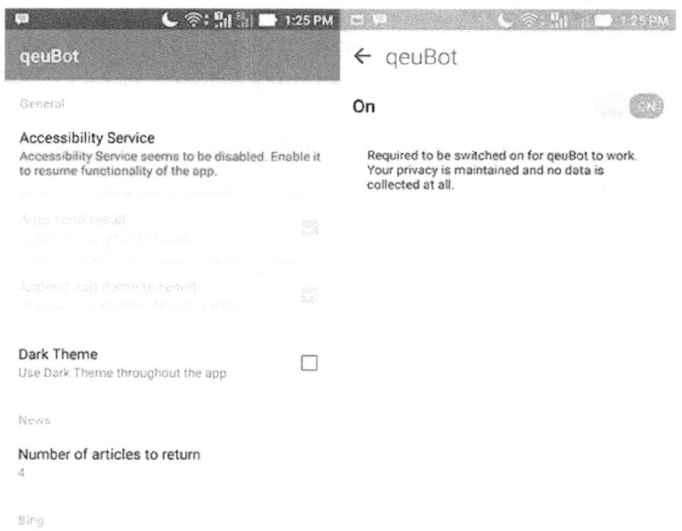

- Once everything is ready, you can begin using WhatsApp Bot by using the appropriate command. For example, find information about WhatsApp on Wikipedia. Just type

WhatsApp @wiki, and the information will be displayed to be sent to the opponent chat.

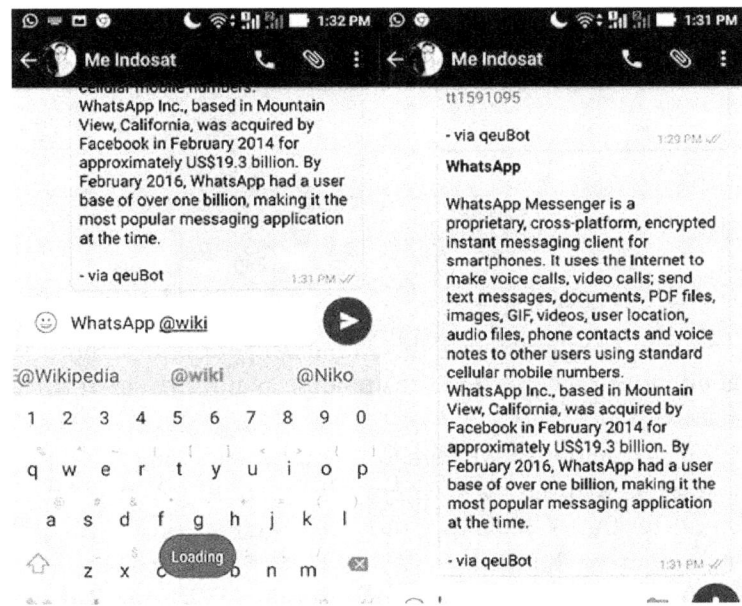

There are several commands that you can use as follows:
1. Command @wiki, ex: WhatsApp @wiki
2. Command @calc, ex: (27/13)-7*11) @calc
3. Command @news, ex: Donald Trump @news
4. Movie title @imdb, ex: Suicide Squad @imdb
5. Command @gif, ex: Baby @gif
6. Command @pic, ex: Mouse @pic
7. Location @forecast, ex: Jakarta @forecast

So, who says Telegram is much better just because having a Bot? You also already can use Bot too for WhatsApp!

4. CHANGE THE WHATSAPP SCENE

Have you ever bored with the WhatsApp default theme which can not be changed? Theme features change is not provided officially by the WhatsApp.

Therefore, we would love to show you how to easy to change the WhatsApp theme without *root*. Simply by installing WhatsApp Plus, you can modify the WhatsApp on your smartphone with a wide selection of attractive themes!

How to Easily Change Theme WhatsApp Without Root
- Before installing **WhatsApp Mod Plus** apk, do not forget to *backup* all your WhatsApp chat beforehand. Go to the **menu settings > Chats> Chat Backup,** then select **Back Up.**

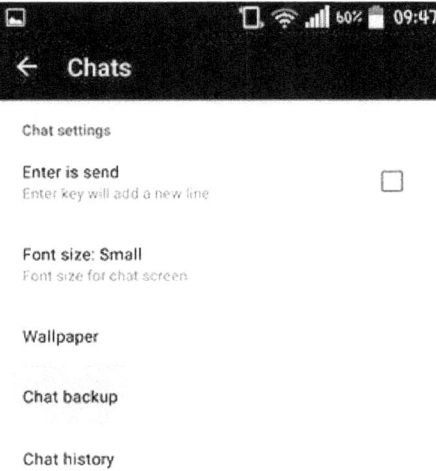

- After **backup** all WhatsApp *chat*, uninstall the original WhatsApp in your smartphone.
- **Install** WhatsApp Plus Mod apk.

Mastering 37 WhatsApp Tricks

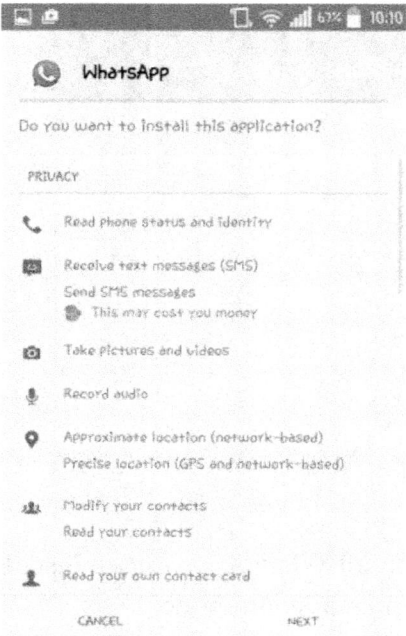

- After you successfully install it, *re-login* and re-use the phone number that you have used before.

- If you want to restore the previous WhatsApp chat history, do not forget to **Restore** your chat backup.

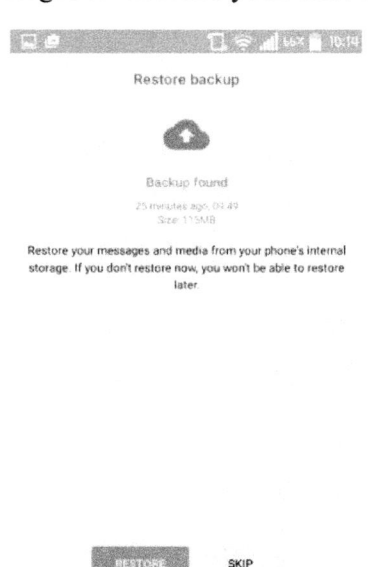

- After completed the stages above, WhatsApp Plus will be ready for use.

Mastering 37 WhatsApp Tricks

- Furthermore, just select any **WhatsApp theme** you want, click **Apply,** and you can use WhatsApp with your favorited theme.

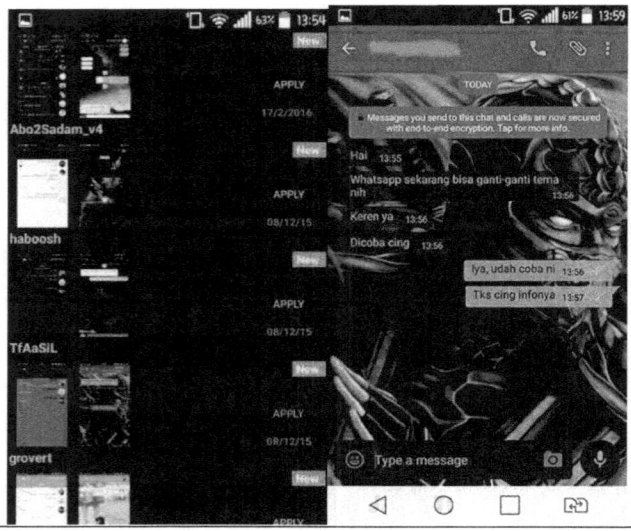

The theme for WhatsApp Plus Mod is very diverse! You can change it whenever you wish. It looks cool, all WhatsApp feature will work perfectly. By modifying WhatsApp, you'll also be able to use the latest features such as WhatsApp video call smoothly.

5. MAKE WHATSAPP ACCOUNT WITHOUT A PHONE NUMBER

With WhatsApp, messaging, audio, and video can be done quickly even if the network connection is unstable. Moreover, the official newspaper of the merger of WhatsApp with Facebook, WhatsApp features now become more powerful.

To create an account, WhatsApp itself needed a phone number. However, what if you want to create a WhatsApp account without a SIM Card? Do not worry, there are WhatsApp tips that you can try to create WhatsApp account without a phone number.

How to Create WhatsApp Without Phone Numbers
- Before using Android WhatsApp tips on this one, you need to install an application called **Primo**. The application is available for many *platforms* such as Android, iOS, Windows, until the MAC.

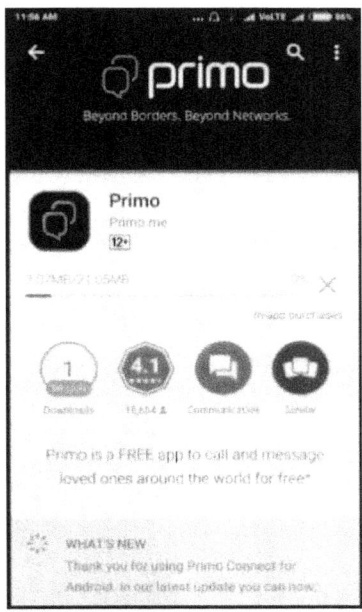

- Open Primo and do the **Sign Up** process for a new account. What we need to do is choose *a username* and *password*.

- Now *logged* in Primo and go to menu, where we will find a cell phone number. Now, use the custom number for account verification WhatsApp.

Mastering 37 WhatsApp Tricks

- Select the method *"as a verification call"* to get a verification code. Wait a few moments and will display an incoming call from the WhatsApp.

- Listen to the code given carefully, record, and enter a numeric code into the verification field.

- Done.

First Step

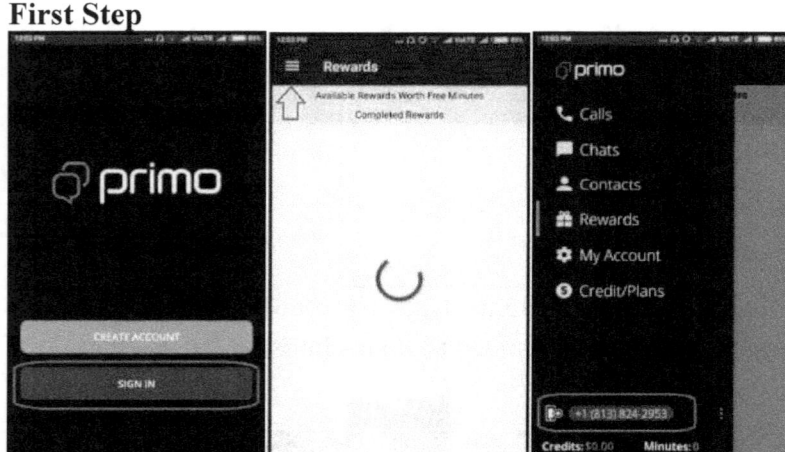

Second Step

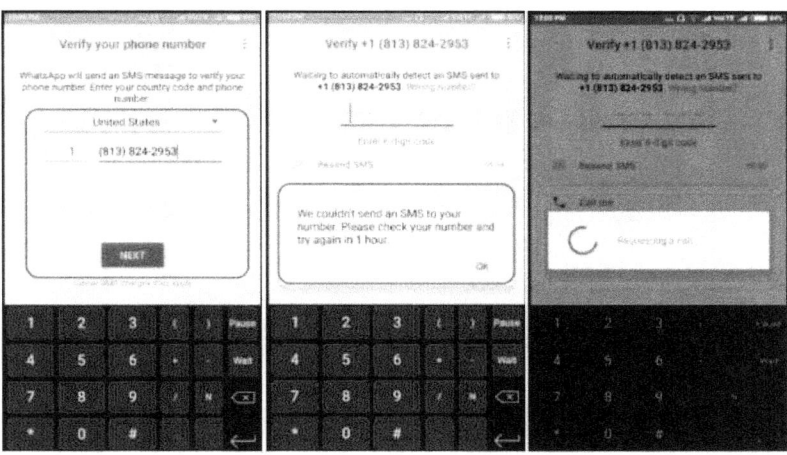

Third Step

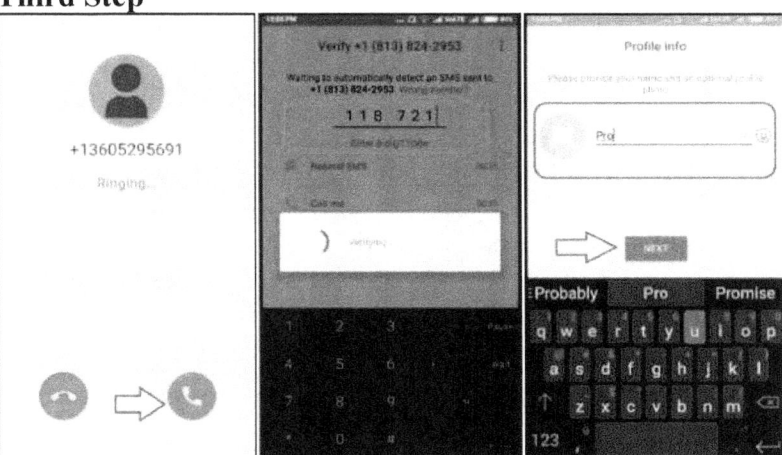

- Now you have successfully performed the WhatsApp account verification without a phone number and now WhatsApp without SIM Card is ready to use!

6. BACKUP WHATSAPP CONVERSATIONS INTO TXT FORMAT

The development of technology, people will be more easy to communicate. Therefore, WhatsApp application is very useful to the user's smartphone. But, if you want to *back-up* conversation WhatsApp how?

Now, therefore we present to tell how *back-up* conversation WhatsApp to TXT format. Could you? Of course, you can! You are probably not yet knowing how. Curious? Let's refer to the following steps:

How to Backup Conversations WhatsApp to TXT Format

To execute it, you must follow the WhatsApp backup conversation steps. To avoid mistakes when putting the backup into TXT format, consider the following steps below.

- **Open** your **WhatsApp application,** then select **Settings.**

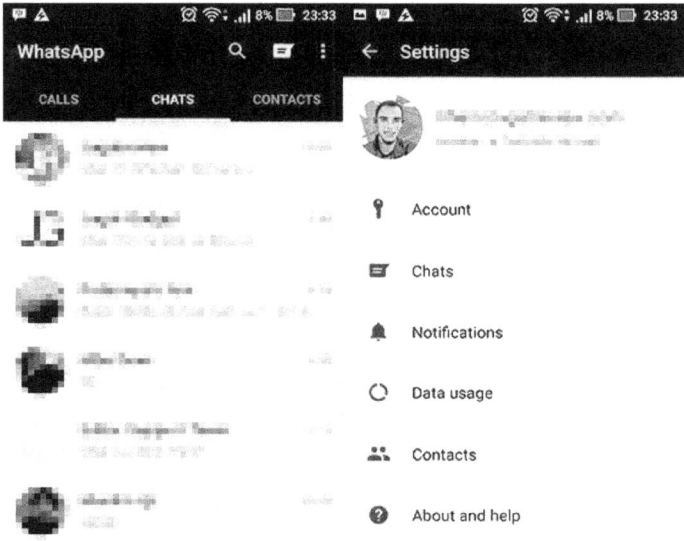

- Then, select **Chats.** Select **Chat History.**

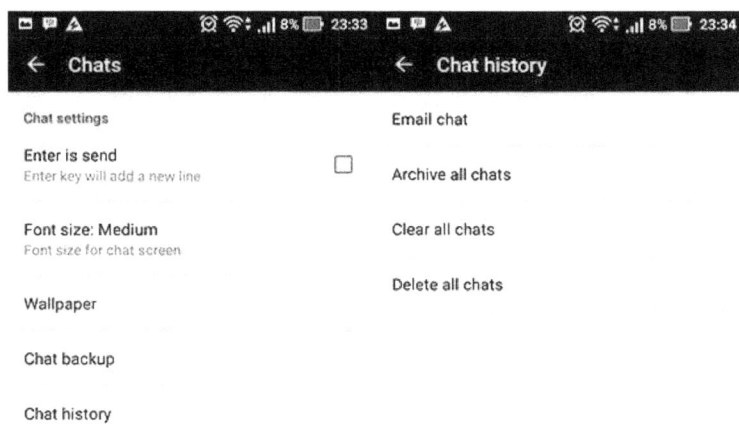

- Next, select **Email Chat** and find the contact you want to backup to TXT format.

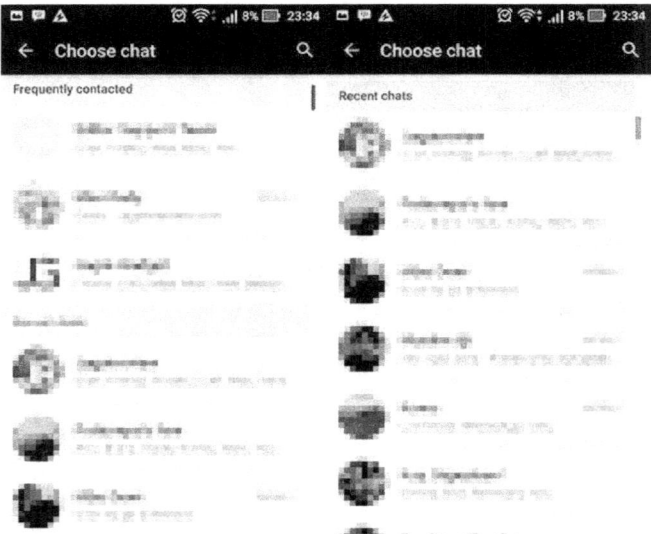

- You will get a *pop-up* **Without Media** and **Attach Media.** Select one. Then, select Gmail to distribute the backup late.

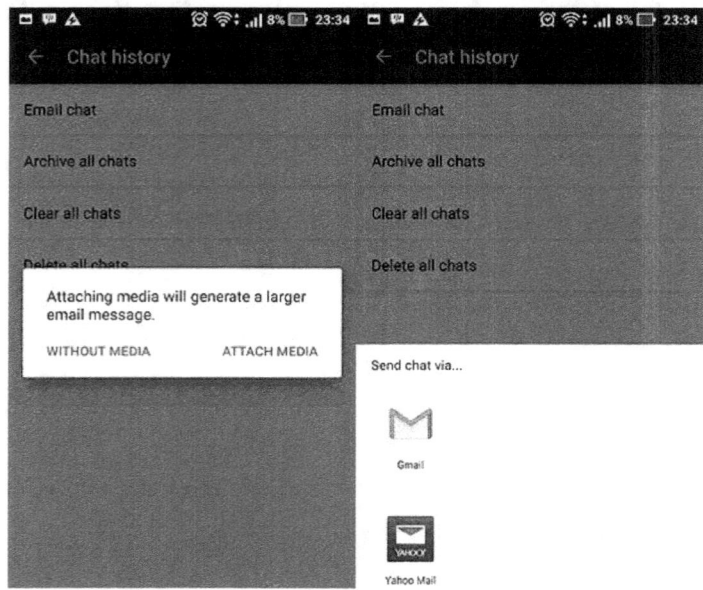

- When incoming **Gmail** format, do not you send it, just keep it in *the draft*.

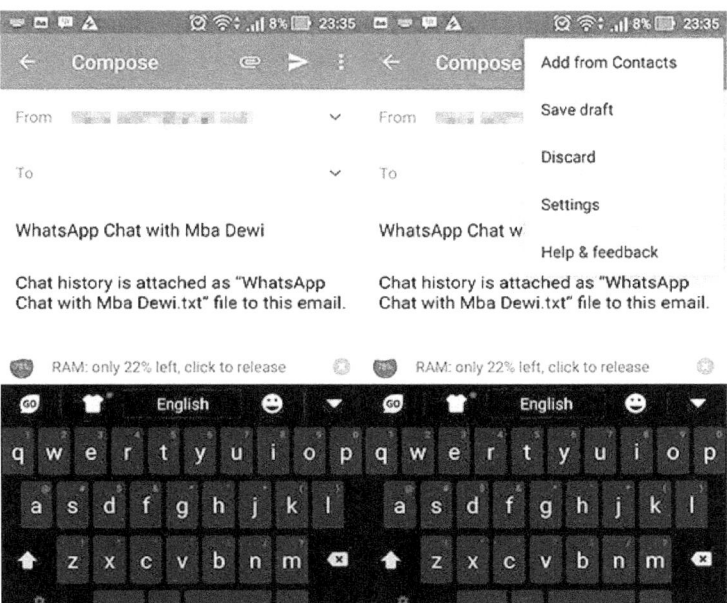

- You enter the draft, then see there is a **TXT file** containing your WhatsApp conversations. Done.

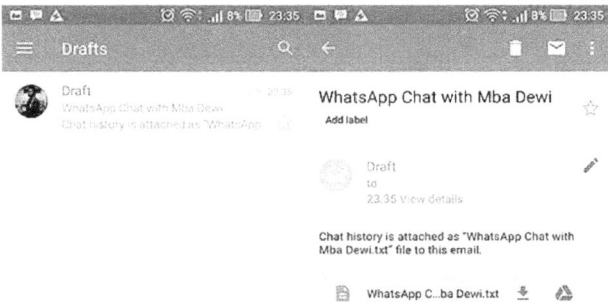

Backup your WhatsApp conversation is very easy, isn't it? Now, you are no longer need to be afraid losing the contents of your important conversations with your colleagues.

7. SAVE INTERNET QUOTA WHILE USING WHATSAPP

Are you an active WhatsApp user? Indeed, active user will consume plenty of data quota. Fortunately, this instant messaging application has a special feature to save data so you can smoothly communicate without worried to the quota limit.

Well, here are the steps that you should set for saving the internet quota while using WhatsApp:

Ways to Save Quota Internet While Using Whatsapp

- Run the application WhatsApp, then *tap* the top right menu bearing the three points and select **Settings.**

- In Settings, please select the **Data Usage** to save quotas when using WhatsApp.

Mastering 37 WhatsApp Tricks

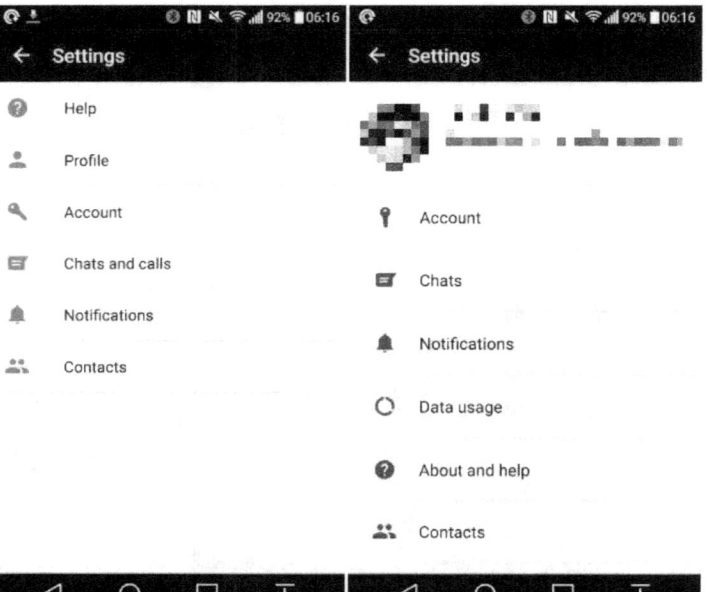

- On the **Media Auto-Download,** there is the option **"When using mobile data".** Here, you can **eliminate all of the checks** so that the smartphone from downloading items such as photos, audio, video and documents automatically when using smartphone internet data. When a message is sent, you will be given the option to **manually download the *attachment*** and choose what files are needed.

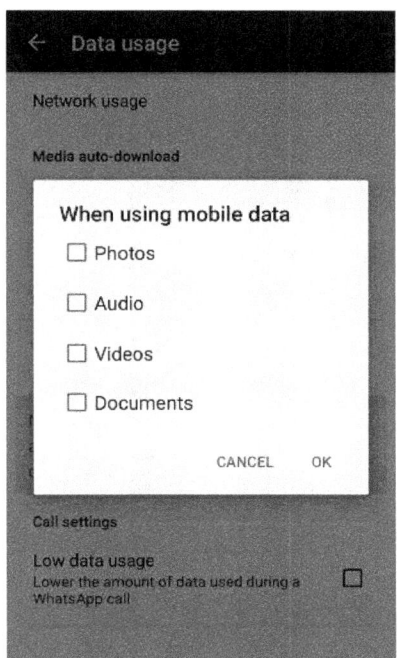

- The next setting, in the section **"When connected on Wi-Fi"**, please check all types of files that can be downloaded automatically when the smartphone is connected to a Wi-Fi network, or you can restrict some file types as desired.

- In the selection of **"When roaming"**, remove all the checkmark.

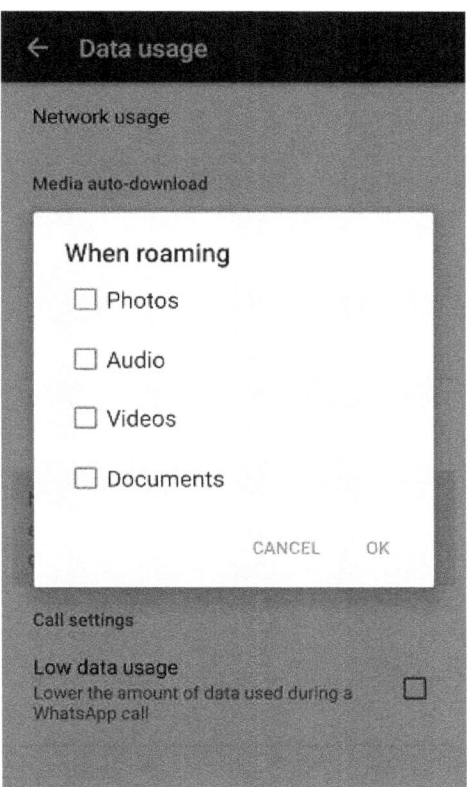

- The final step, please check on the option **Low Data Usage.** By checking this option, the use of the data will be suppressed when you make a phone call or video on WhatsApp thus saving the internet quota.

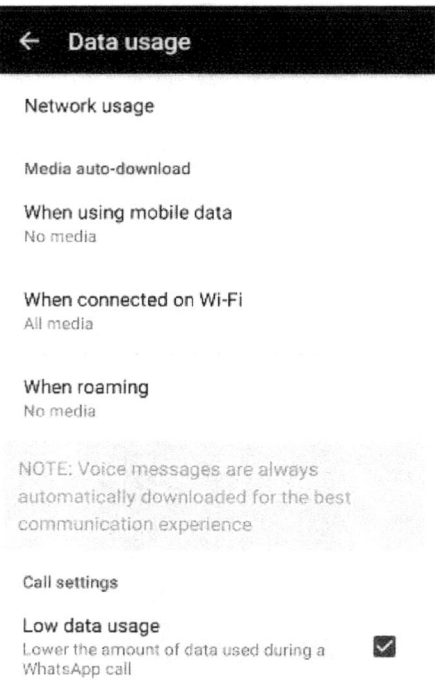

- By saving quotas as explained above, the data usage of WhatsApp application will be trimmed. You can still save your pocket when using a data access from smartphones. It's easy, right?

8. TURN OFF THE MARK "READ" BLUE TICK

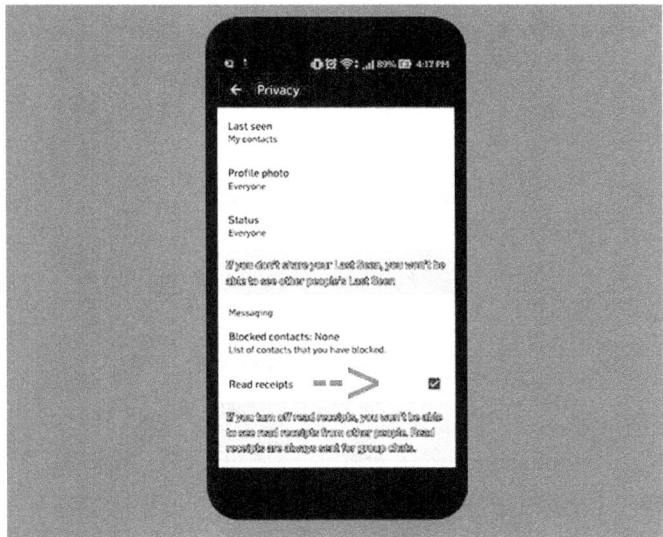

Any messages sent and read will be marked with two blue tick. If you have objections interlocutors know the status of the message has been read, you can change it.

Go to the menu **Settings> Accounts> Privacy.** Next step, scroll down and uncheck **the Read Receipts** menu option to disable it.

9. HIDING STATUS "LAST SEEN"

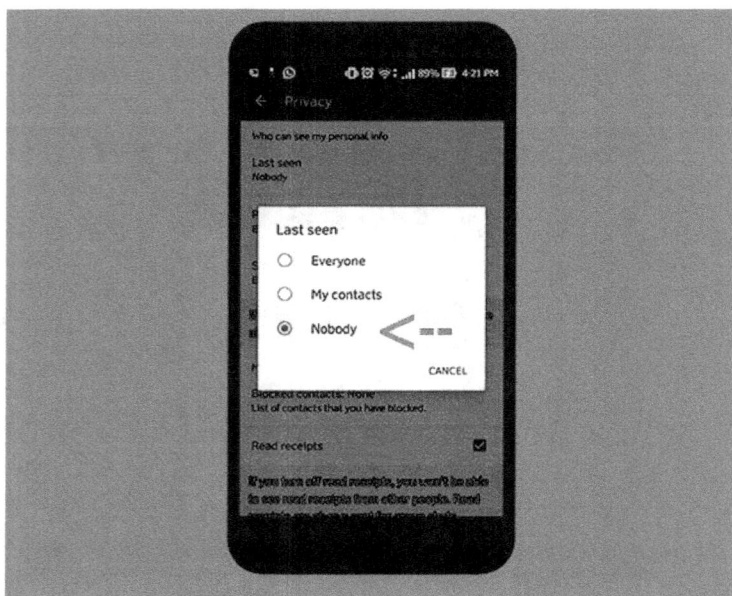

Sometimes we do not want to reply to a message from someone, but because WhatsApp allows users to see each other their last online time, we are uncomfortable with this situation. When this is considered as entering your privacy, you can hide the **"Last Seen"** status.

The trick, go to **Settings> Accounts> Privacy.** Then select menu "Last Seen", it would appear another option in *the pop-up.* You simply need to choose **Nobody,** so no one else could see the information.

10. KNOWING YOUR MESSAGE HAS BEEN READ WHILE BLUE TICK IS BEING TURNED OFF

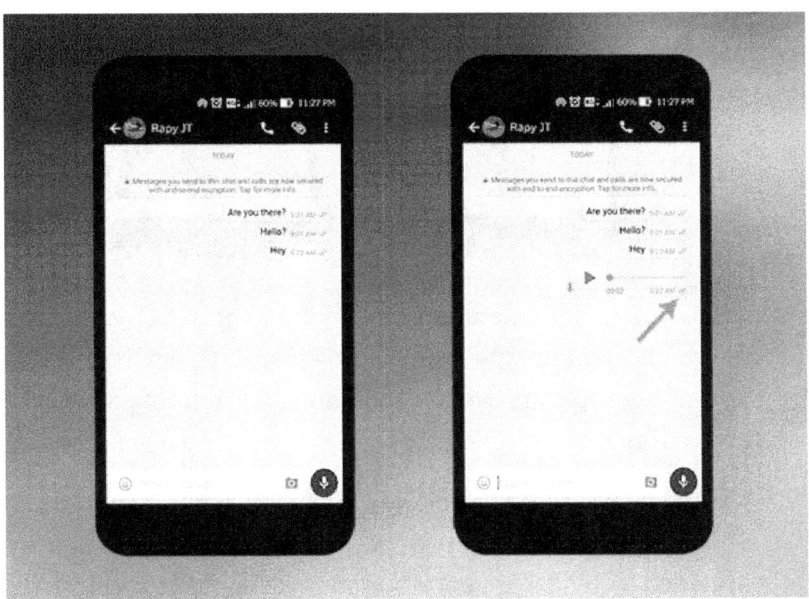

Did you ever having trouble contacting friends who disable the Read Receipts feature or blue tick? Not to mention, the feature "Last Seen" was also turned off.

To be sure, we have a new trick to know your message was already read or not even when the blue tick was turned off. The trick is to send a voice recording.

When you send a voice message and your friend play the message. Then two blue tick will still appear no matter if the Read Receipts feature is being turned off.

You will notice whether she was deliberately ignoring you or not. Moreover, there is still another way.

First step

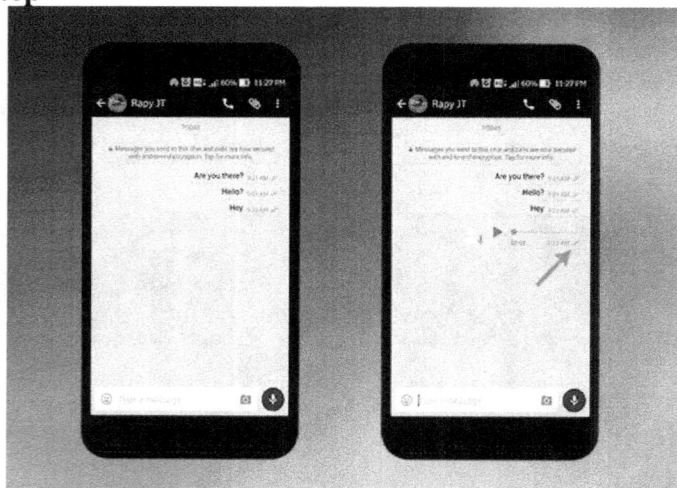

1. This trick is documented in the WhatsApp FAQ, but it is not much known to many people. To ensure if your friend read your WhatsApp messages, the way is by sending a **voice recording.**

2. Maybe he read every message you send. So, when you send a voice message and your friend open the message. Then two blue tick will still appear no matter if the Read receipts feature is being turned off.

Both ways

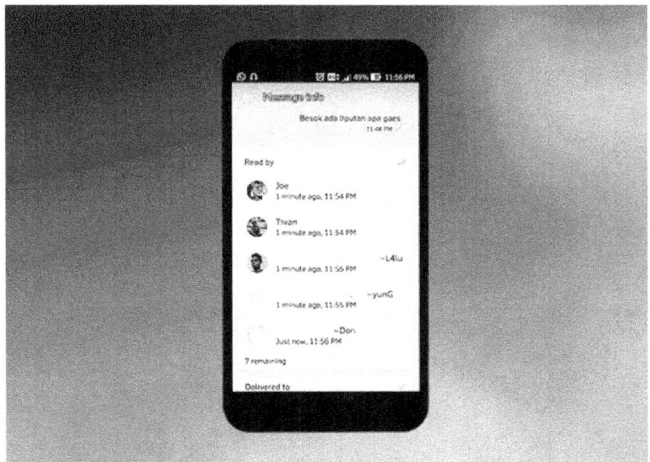

1. You need to know, read receipts feature only applies to private messages only. A read-receipt for all sent messages in the conversation group (group chat) is always active.

2. So, no need to send a voice message to a *group chat*. Just type a word or phrase to the group which your friends also joined there. You can see who has viewed your message, including your friend. If he read in a group *chat*, that is mean she was deliberately ignoring *personal chat* from you.

3. That's the way you know the message is being read even a blue tick was off. Frankly, WhatsApp already presented the features, of course, to pamper users. The rest is how you as a user adjust it to satisfied your needs.

11. EDIT PHOTOS ALA SNAPCHAT BEFORE SEND

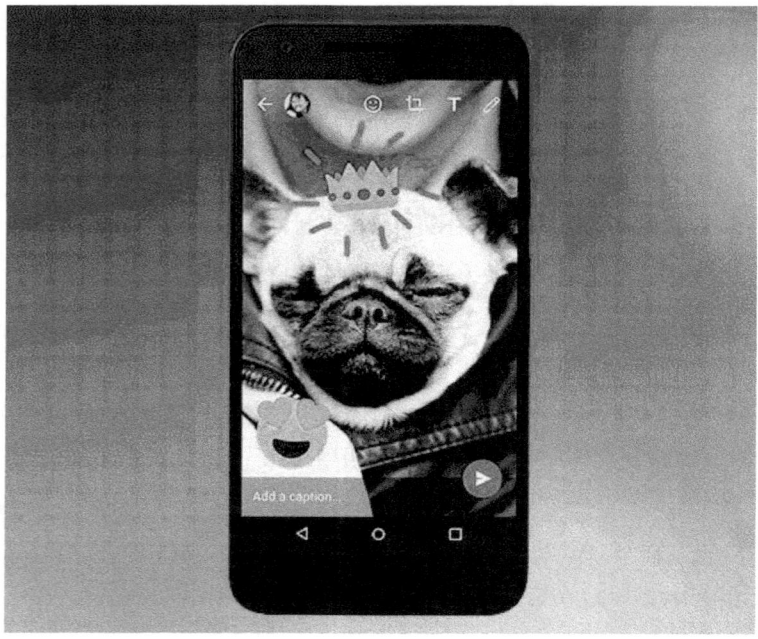

WhatsApp has added *drawing* features and the ability to add a sticker on WhatsApp camera as well as on Snapchat. In fact, the interface looks very similar.

So, after you take a photo or video. You can add emoji, scribbling, or add text, just like Snapchat.

In addition, WhatsApp also added *flash* features in front of the camera which illuminates an object so that the result is brighter when *selfie* in a minimum lighting. WhatsApp latest camera features on this matter should have been rolling on the Android today. What do you think about Facebook blatantly plagiarized Snapchat?

12. GIVE STAR SIGNS

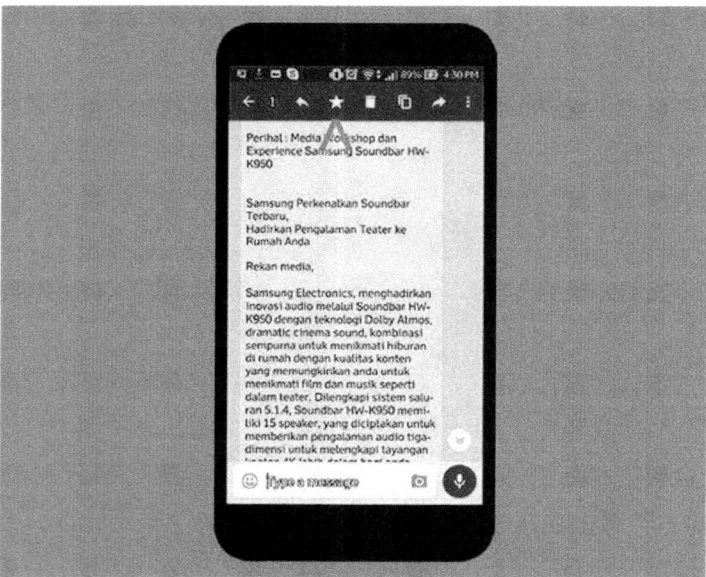

Is there any important information in a conversation? Don't let it sink and disappear, you can give it a star sign. This sign is useful to keep track of an important information.

The method is simply select the message you want to mark, tap and hold for a few seconds. It will appear the star in the WhatsApp *header*, right next to the icon "delete", and then select the asterisk.

When you want to view the messages that have been marked with an asterisk, you just touch the three-dot menu at the upper right corner. Next, select the tab marked **"Starred Messages"**.

13. MENTION FRIENDS IN GROUP

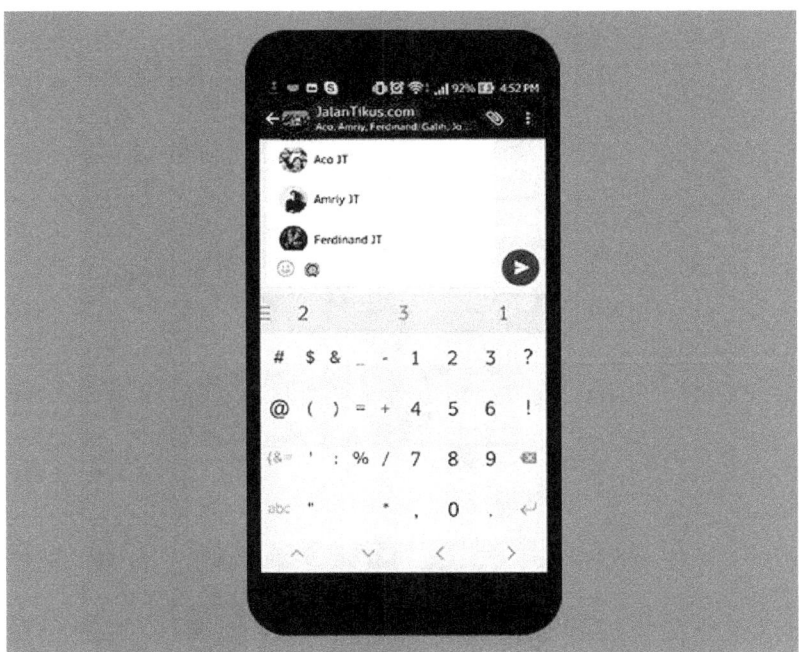

Like Twitter, WhatsApp also got a *mention* feature in the group. Using this feature, members of the group would be more aware while they were *mentioned*.

Mention feature can be used by typing the character "*@*". After typing the symbol "@" in the conversation column, a row of contacts in the group will be shown.

14. REPLYING TO A MESSAGE DROWNED IN GROUP

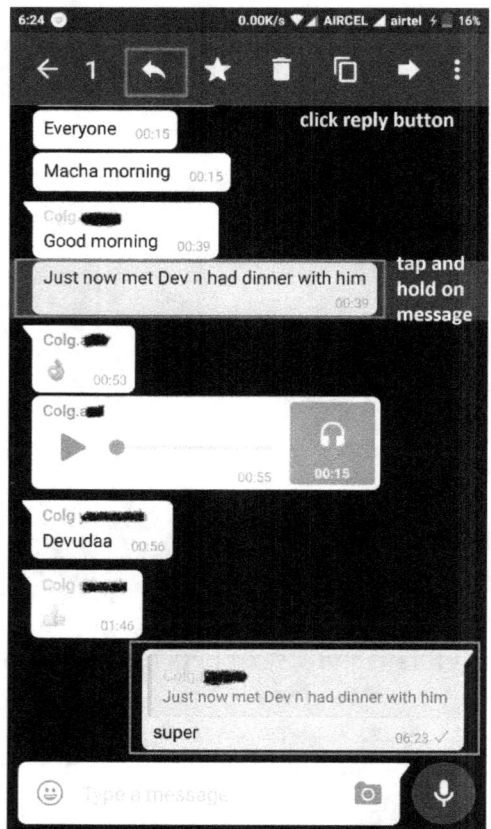

Because of the height of the conversation in a group, sometimes we are often confused. Which question did our friend answer?

Let's make it neater, you can quote one of the messages in the group and reply it. Tap and hold on the message, then WhatsApp will show some icons to act.

One of them, on the far left, there is **an arrow icon.** You just press it, then WhatsApp will automatically quote the message you want to comment on. And finally, it's worth noting that you can quote any participant in a chat, including yourself, and that media messages show up along with their preview inside the quote. That's sweet.

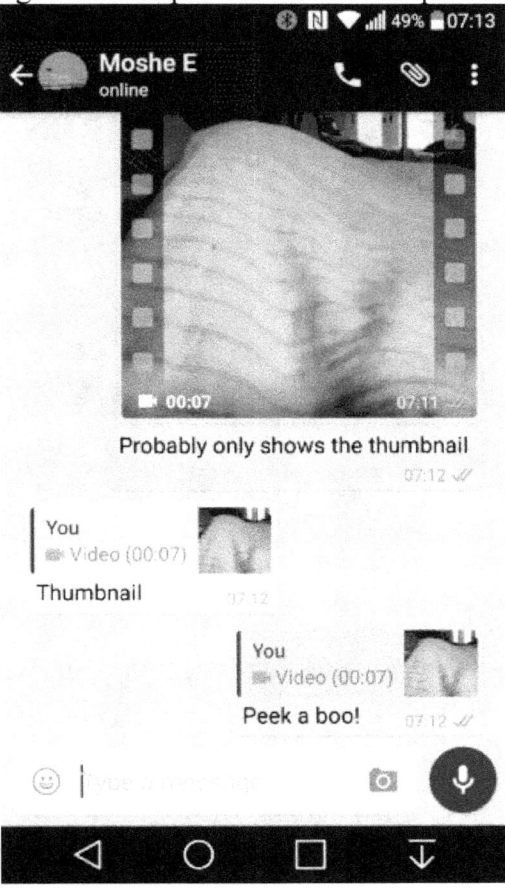

15. TURN OFF AUTOMATIC DOWNLOAD

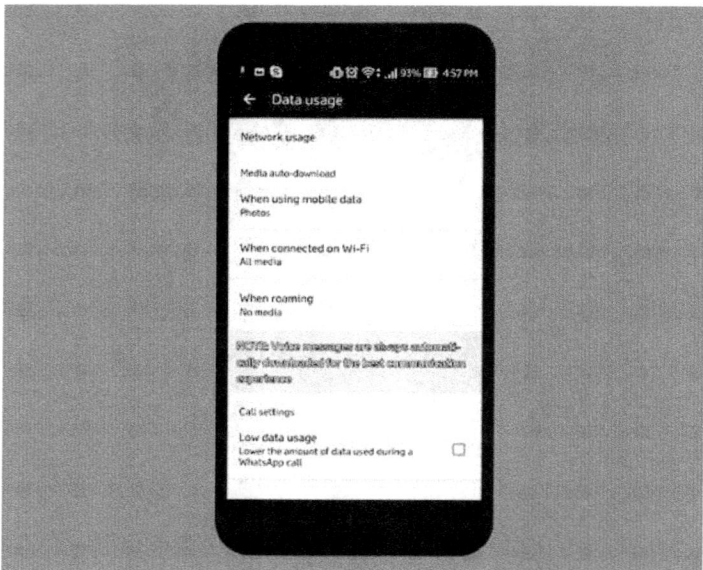

Your internet quota package runs out? You can opt to no media auto-download in the settings so that all the media should manually be downloaded.

Commonly, photos and media that is sent through WhatsApp will be automatically downloaded. This, of course, runs the quota consumption.

How to manage it? Go to the **Settings** menu, and select **Data Usage.** There is a *tab* labelled **Media Auto Download** that contains the settings for the automatic download.

16. BLOCKING CONTACTS

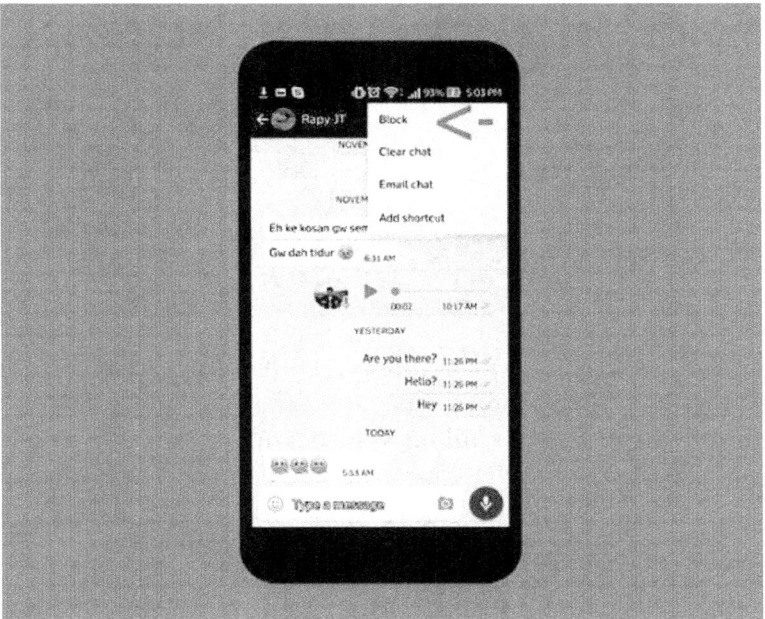

Have you often get a message from a stranger out of your friend list? Simply block it. WhatsApp provides this feature and it can be easily accessed.

The trick, open conversations with people who are targeted. Touch the three-dot menu at the top right and select the **More** tab. Then you will found the **Block** option among the menu.

17. VIDEO CALL

After a long wait, finally, WhatsApp released his *video call* feature. Unfortunately, not all WhatsApp users can try this feature. **How to Video Call on WhatsApp**
- Go to user **profile** who would you call and **press** the **Call button.** Unlike the old version, you can choose to make a *video call* or a *voice call.*

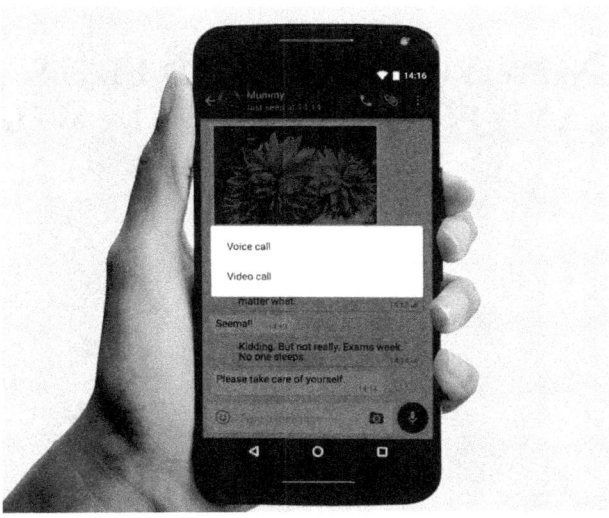

- Here is a *video* display on WhatsApp *call*.

Interesting, right? Good luck!

18. NOW YOU CAN SEND FILES WORD, EXCEL AND POWERPOINT VIA WHATSAPP

When sharing a document feature first announced, we can only send files with PDF. WhatsApp is now increasing its support to send documents with file extensions wider coverage. You can send files in Word, Excel, and PowerPoint extension. So how to send the document?

1. Open a chat.

2. Tap the icon at the top of the screen.

3. Choose Document.

4. Select the desired document to send and tap Send in the popup.

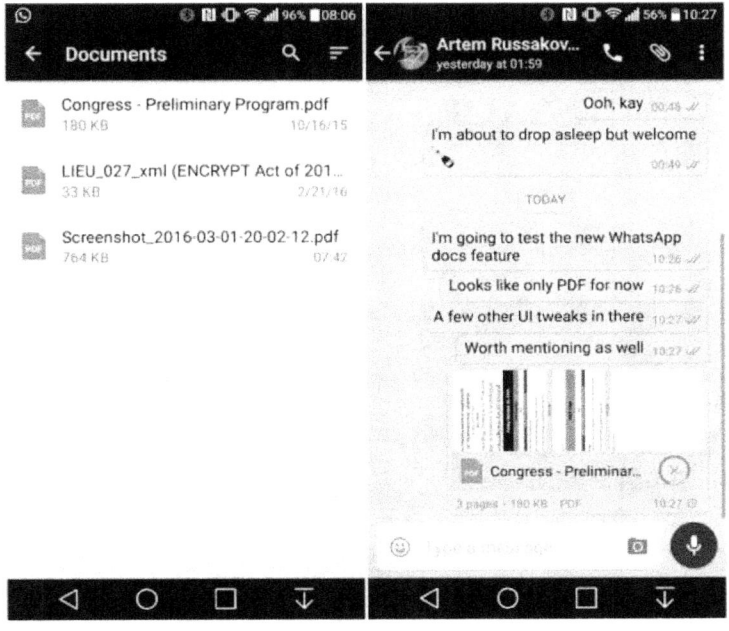

19. AUTOMATIC ENCRYPTED MESSAGES WHATSAPP

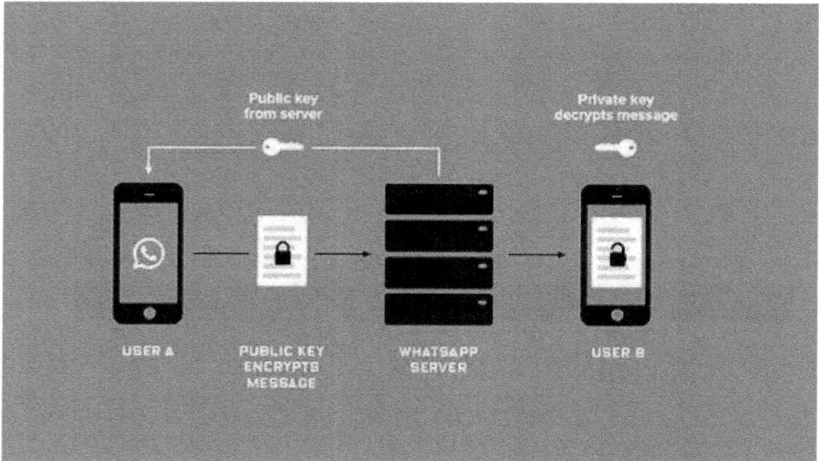

WhatsApp messaging service reaffirmed its commitment to maintain the privacy and security of user data by encrypting the conversation between users. Thus, all the messages, photos, videos, files, and voice messages you sent are automatically encrypted *end-to-end*. In fact, including group chat and voice calls.

The idea of this feature is quite simple. When users send the message each other, then the only person who can read is our interlocutors or the people in the group. No one else can see the message, no *cyber crime,* no hacker, even though the WhatsApp parties.

20. QUICK REPLY NOTIFICATIONS

Quick Reply feature will help you to reply a message quickly and easily without having to open the WhatsApp. This Quick Reply works on any screen, including in the *lock screen* mode like *pop-up* notifications features. After tapping the *reply button,* you will find an *overlay* where you can type and send a reply without having to open the application.

21. CAN SEND TEXT BOLD AND ITALIC

You can tamper with typed text, to makes the text **bold** or *italic*. To note, the text formatting is not obtained from a button. You must enter a unique code so that the text is slightly changed. Now to add Bold formatting, just add (asterisk) "*" before and after the text. For example, ***Hola!***. Hit the **Send** button, and WhatsApp will automatically make it bold.

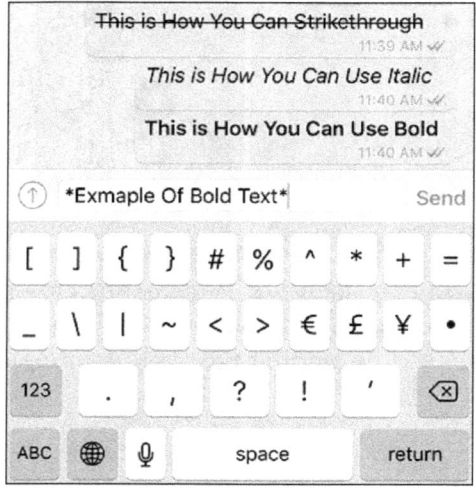

To type in Italics, you'll need to add (underscore) "_" before and after the text. Same like Bold formatting, WhatsApp will now convert the text to Italics.

The last and the final is Strikethrough. To use that, you'll need to add (tilde) "~" before and after the text.

That's all folks! It isn't a great deal for a company like WhatsApp to deliver these basic formatting features. Anyways, it's rightly said, better late than never.

22. RETRIEVE DELETED MESSAGES

Most of the smartphones users using WhatsApp as the main *chat* application. Including you, right? Surely you have such message conversations that are very important, but what if you deleted or missing this conversation? Either because you change your mobile or loss of the device and other causes.

Data Backup WhatsApp

First of all, you should make sure that backups were already activated. **Local backup** is automatically performed at 02.00 a.m. each day and your *database* will be stored in a file in the phone memory. In addition, you can also back up through Google Drive.

Backup to Google Drive

You can back up your messages and media images, video, or sound to Google Drive. Do you have a google account? If you do not have a Google account already set up, tap Add account when prompted. So, if you lose your phone or replace the new device, *chat* history remains safe. Here's how:
- Open WhatsApp.

- Select button **Menu> Settings> Chat and Call> Backup chat.**

- Click on Backup to Google Drive and select the backup frequency that you want, or click on **Backup** for immediate backup.

- You will be asked to choose a Google account to use for backing up your WhatsApp *chat* history.

- Click on **Back up** through to select the network you want to use to perform the backup.

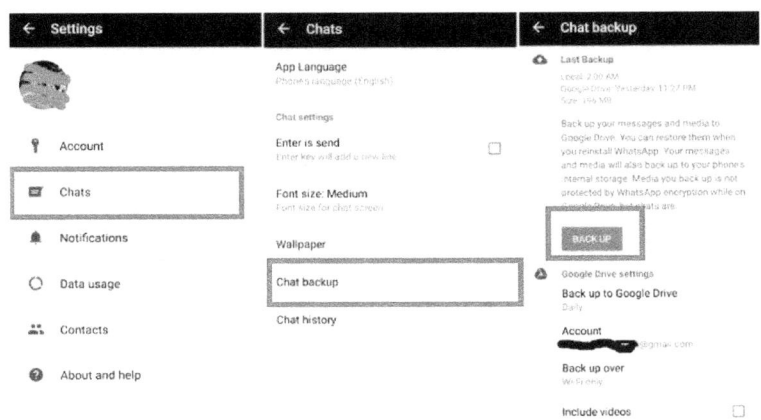

How to Recover WhatsApp message from Google Drive

Mastering 37 WhatsApp Tricks

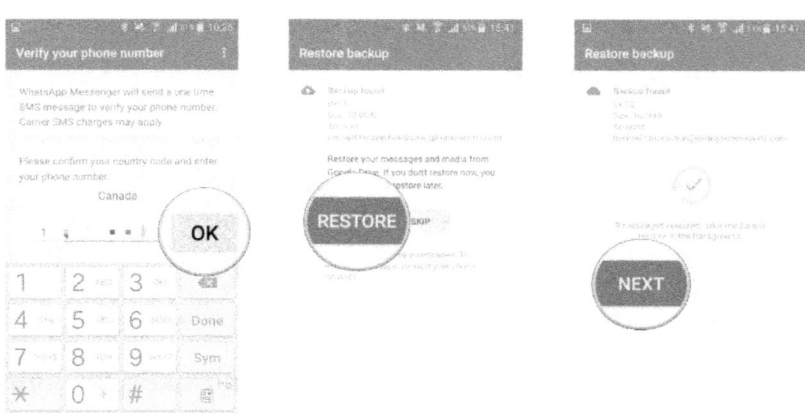

After you install WhatsApp and verify your phone number, **WhatsApp** will offer you to restore *your chat* history and media from the backup you'll ever make. Select **"Restore"** when asked to do so.

If WhatsApp can not detect the backup, it is likely because you may not be connected to the same Google account. It could be also because you do not use the same phone number used to make a backup.

How to Restore Old WhatsApp messages in Phone Memory

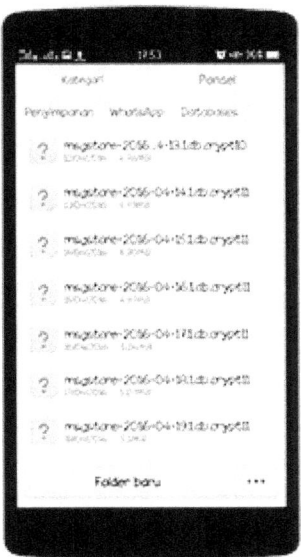

If Google Drive will only keep the most recent backup, **WhatsApp** saves a backup in the phone memory in the last 7 days. Here's how:

- Open **File Manager** application.

- Go to **sd card / WhatsApp / Databases.**

- If the data is not stored on the SD card, there may be internal storage

- Call the backup file you want to recover from **msgstore-YYYY-MM-DD.1.db.crypt8** to **msgstore.db.crypt8.**

- Clear app WhatsApp.

- Replace the application WhatsApp.

- Choose a restore when prompted.

Mastering 37 WhatsApp Tricks

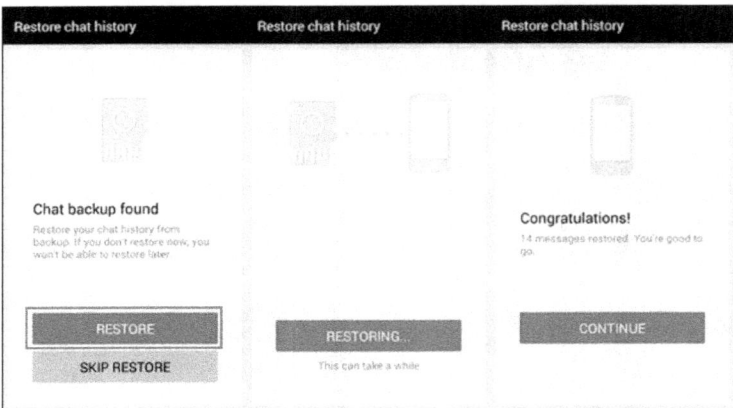

This all trick ensures all WhatsApp messages and media will be safe, even if you lose your phone or change the device.

23. KNOWING SOMEONE LOCATION THROUGH WHATSAPP

For some reason, if you want to know a person's location via WhatsApp messages or Facebook message, it can be tracked. For example, your girlfriend said she is at home. But you doubt because per information from a close friend, she goes out of town. You can try some cool tips to got her location. Here is **how to determine a person's location via WhatsApp.**

Tracing the IP Address Using the Command Prompt

You know, using a feature **command prompt** on your computer can trace the IP address of the person you're *chatting* on WhatsApp. The trick is to follow the steps below.

Mastering 37 WhatsApp Tricks

- First, you open the **web WhatsApp,** after which it started to *chat* with the target to obtain an IP address.

- Make sure all the applications currently running in the background has been stopped. Press **Ctrl + Alt + Delete** to open the **task manager**. Close all application except for the browser that you are using to *chat*.

- Now press **Win + R** on your computer keyboard.

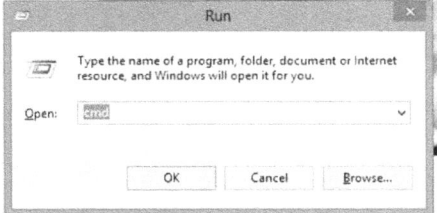

- Type **cmd** and press **enter**.
- At the **command prompt** that appears type **netstat -an** and press **enter**.

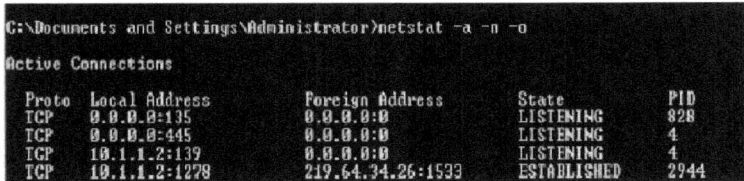

- Write down the IP address listed.
- *Tracking* to know the exact location via the website **http://www.ip-adress.com/ip_tracer/**.

This method also applies to determine a person's location through Facebook chat. Interested in trying a way to know the location via WhatsApp? Do not misuse it.

24. READ MESSAGES WHATSAPP WITHOUT ELIMINATING BLUE TICK

Maybe you ever annoyed because of your privacy to ignore someone message was being disrupted. Imagine, now people know whether you've read the message. In the end, so you were being clumsy to simply ignore him?

How to Read WhatsApp Without Blue Tick

- ✓ message successfully sent.
- ✓✓ message successfully delivered to the recipient's phone.
- ✓✓ the recipient has read your message.

1. Peering Quick Settings

Do you have WhatsApp incoming messages from people you are avoiding? Maybe from an ex or a salesman. To avoid the appearance of a blue mark on the messages you read, you can read the incoming messages from Quick Settings. WhatsApp messages read from the Quick Settings or Notification Bar will not turn the read status to blue tick mark.

2. Using the Widget

In addition, to enhance the look of your Android smartphone *home screen,* **WhatsApp Widget** has another function that can be used to read WhatsApp messages undetected. As long as you read the incoming message in the widget, then a blue tick mark on the sender will not appear. But, once you reply, blue tick mark will show.

3. Airplane Mode Method

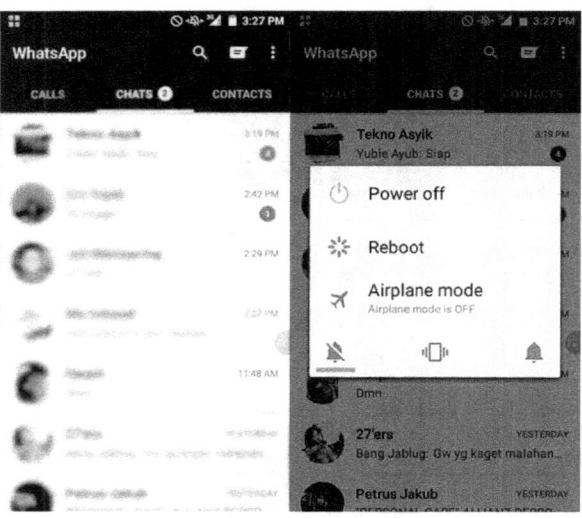

You may be upset if your message has been read, but had no replied? Yes, you will be going to feel ignored. Well, to avoid hurting her feelings, you can do it without any 3rd party application. To read WhatsApp messages without giving a blue tick on the sender, you can enter **Airplane Mode** when messages arrive. Then read the message. Once read, you can go out Airplane Mode again. Thus, *chat* on the sender will not turn into a blue tick. It's easy, right? But this only applies to the iPhone users, while Android does not apply.

4. Privacy setting in WhatsApp

While entering Airplane Mode for each incoming *chat* is not practical, you can use other means provided by WhatsApp. Simply by going into the WhatsApp settings, and select **Account.** Upon entry the Account settings, select **Privacy,** then change **Last seen** into **Nobody**, and remove the tick in the **Read receipts.**

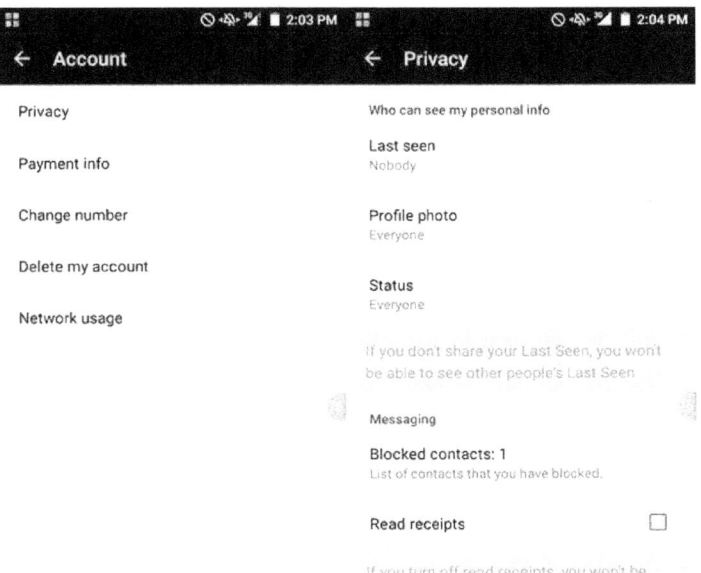

Furthermore, each *chat* you read would not turn the mark into blue in recipient's interface. Unfortunately, you would not know either if the recipient has read the message.

5. Use WhatsApp Web

Having launched by the end of 2015, **WhatsApp Web** already supporting most web browsers. Well, by using the Mozilla browser, you can hide the blue sign in the *chat* you read in WhatsApp.

Simply **install the Add-on ShutApp** in your Mozilla browser. This add-on can only be used in the Mozilla browser.

Mastering 37 WhatsApp Tricks

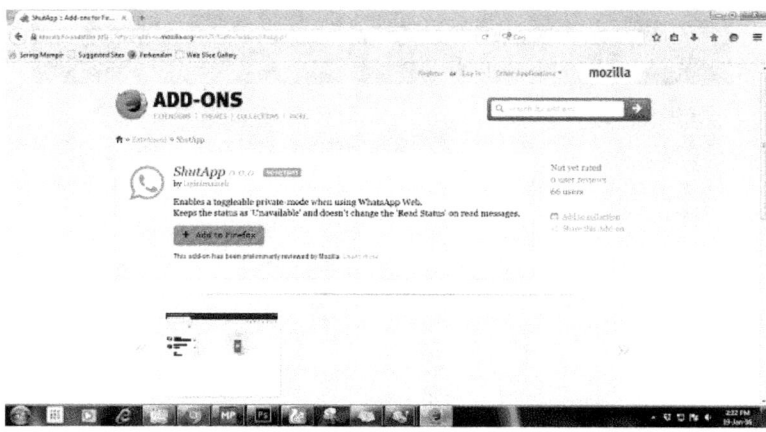

To use it, click the WhatsApp icon on the *toolbar* to activate ShutApp when you're using WhatsApp Web. Once clicked, you will go directly to **WhatsApp Privacy Mode.** Click again to end the session if you no longer want to hide the read status in the WhatsApp sender.

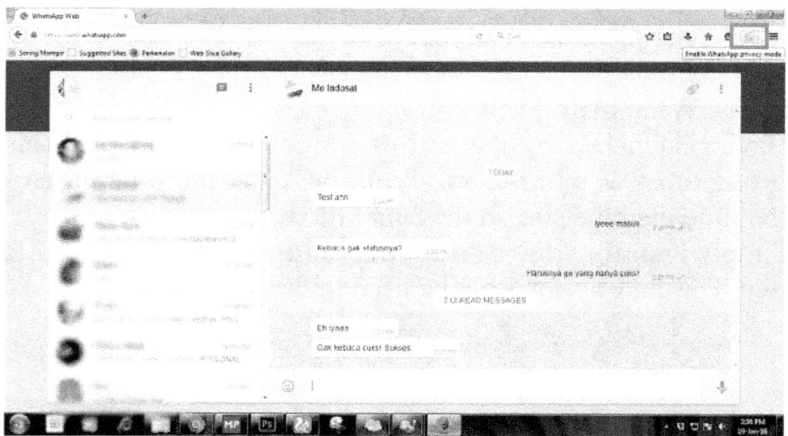

6. Using WhatsApp Plus

As the name implies, **WhatsApp Plus** equipped with myriad advantages over **WhatsApp Official.** One of the interesting features of WhatsApp Plus is the function to hide your *online* status, blue tick, when messages have been read, the two tick when the message is read, etc. So, you can easily hide your status with WhatsApp Plus.

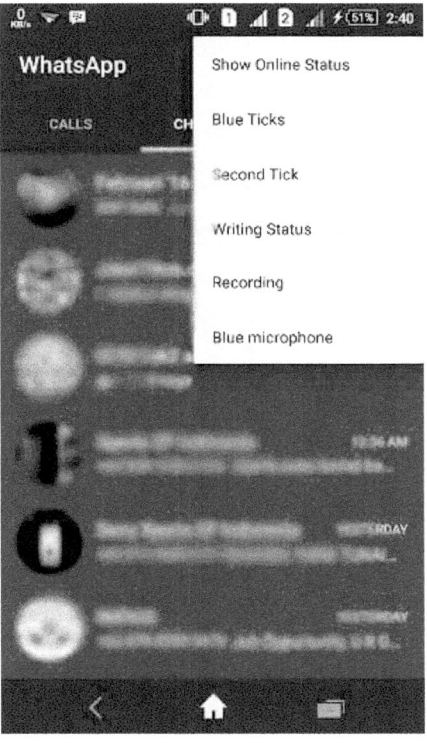

Keep in mind, as a suggestion, do not hide the two tick checklist. Because we had experienced that the incoming message will be slow.

25. WHATSAPP ENABLES USERS TO CHECK THE MESSAGE READ TIME

Sometimes you just send hello messages to our friends or sorry messages for forgiveness. You hope for replies but none. It will make you embarrassed or upset. In fact, it is easy to check if people have read your messages.

Just go to one of your WhatsApp chats, and hold on one of the messages, then you will see the Info option (circle i), tap and a new window will show you the exact time of delivered and read messages. You may also try this trick in your group messages.

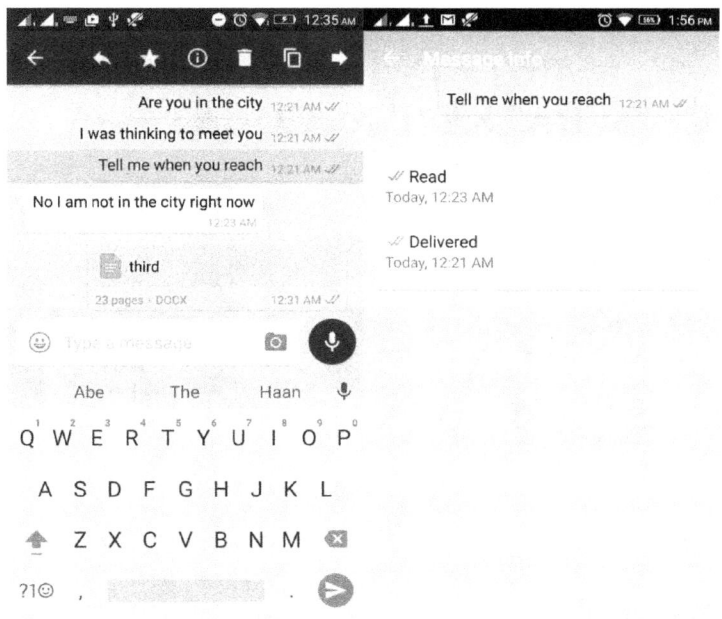

26. USING 2 IN 1 SMARTPHONE WHATSAPP ACCOUNT

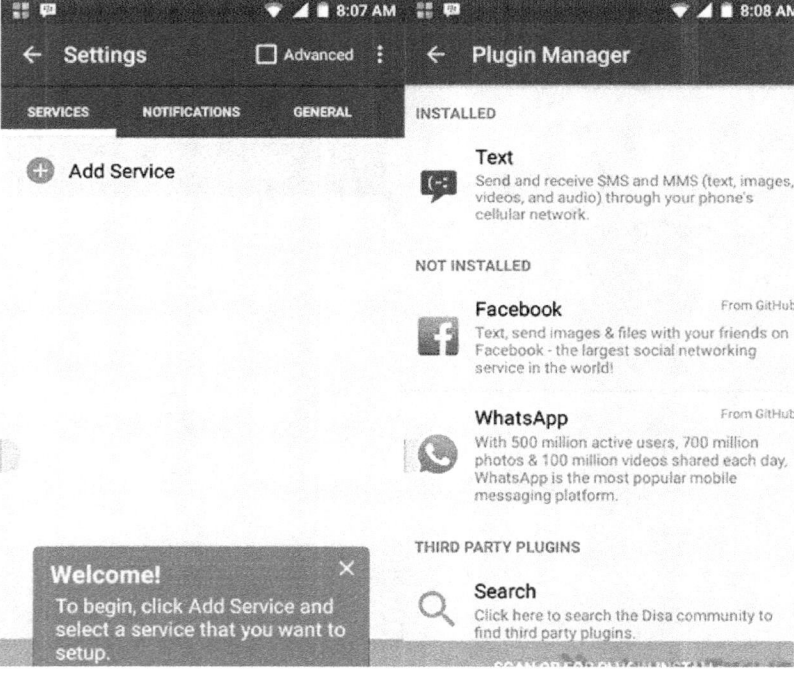

Even if you use a dual SIM smartphone, you are only able to use one number. However, there is a trick to use two WhatsApp in a single smartphone. Simply use an application called **Disa**. By using 2 WhatsApp in 1 Android smartphone, so you can divide which WhatsApp specifically for families and a friend, and which one for work activity without having to carry a lot of smartphones. Do not get bothered by WhatsApp for your company work during the holidays. Especially in a long holiday by the end of the year.

How to Install 2 Number WhatsApp in One Android device

There are a lot of applications you can use to create 2 WhatsApp in 1 smartphone, like OGWhatsApp or WhatsApp Plus. But commonly these applications have certain limitation. Using several tips below will guarantee you can use it easily.

Problems Using two WhatsApps Number in one device

WhatsApp Plus and **OGWhatsApp** apk are popular applications that can be used to install 2 WhatsApps in 1 smartphone. However, due to the WhatsApp security, these applications always run into problems. The problem is usually the expiration of the WhatsApp application. As you know, WhatsApp subscriber should pay annual charge. Another issue is failed installation since the WhatsApp primary application was already being installed.

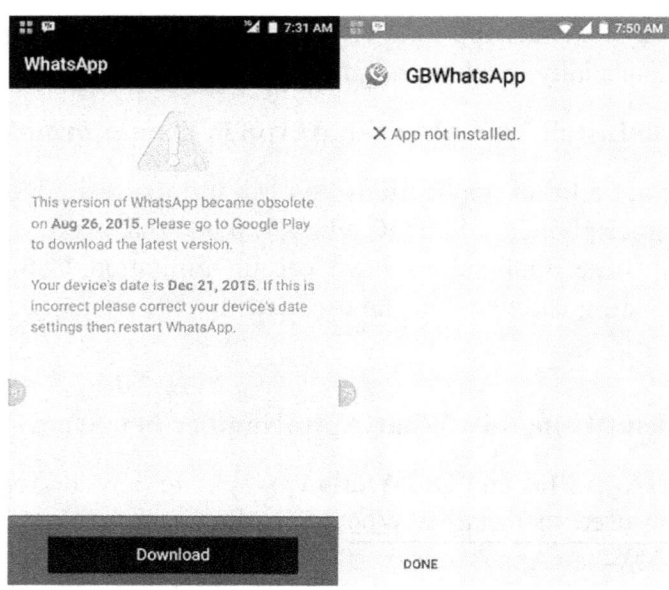

How to Install WhatsApp Android 2 In 1 By Disa

Because these applications are always accompanied with complicated problems and difficult installation, we offer another way to install 2 WhatsApp in 1 device. Disa helps you to install two WhatsApp account in one Android smartphone. Disa can be downloaded from the google play.

Disa 0.9.4

Disa can be used not only to combine a variety of incoming messages in one place, but also has functions to use 2 WhatsApp account in one Android smartphone. The installation of Disa is as follows.

- Once installed, open Disa. Just continue *Swipe* to skip *the tutorial* section, and select agree to its **Terms of Service.**

- Select **Add Service.** Then in the **Plugin Manager** select **WhatsApp.**

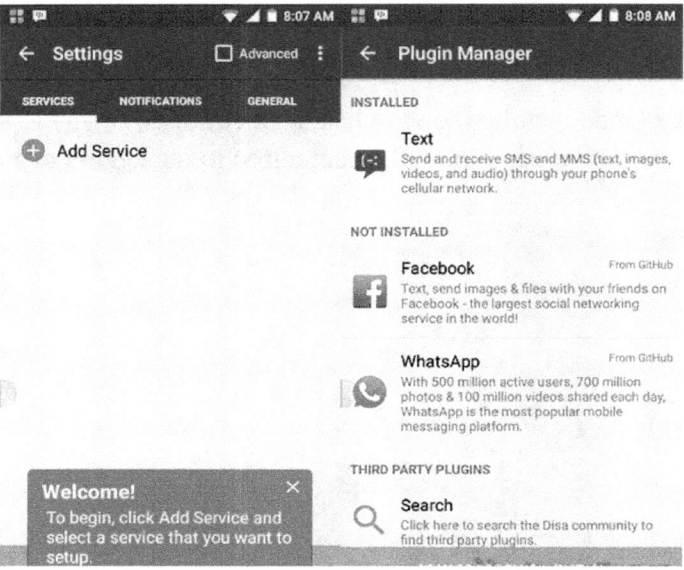

- Wait until the *plugin* installation is complete. It takes only 2-3 seconds. Then **Restart** Disa application.

- After a restart, you will see the **WhatsApp Setting Needed** interface. Please click on it, then Skip it when you see **WhatsApp Setup Wizard.** We can ignore it because we will use another number for the second WhatsApp account.

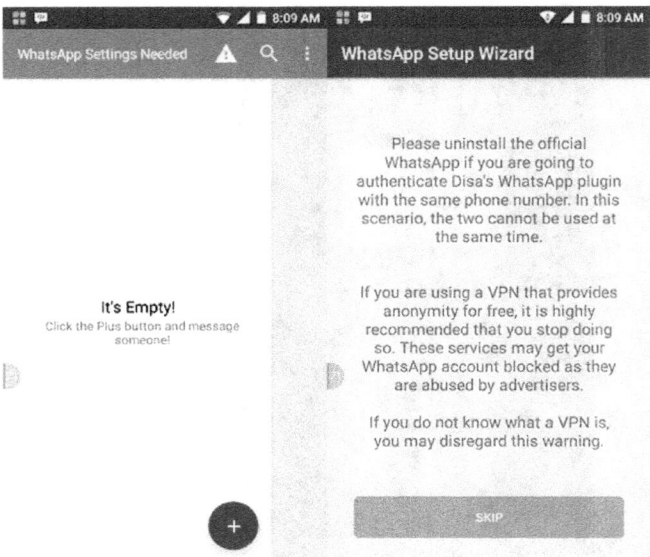

- Please enter your name and number, then tap **Next**. Just ignore the numbers 510 and 010 parts of it. Please confirm the number, then wait for incoming SMS verification. Done.

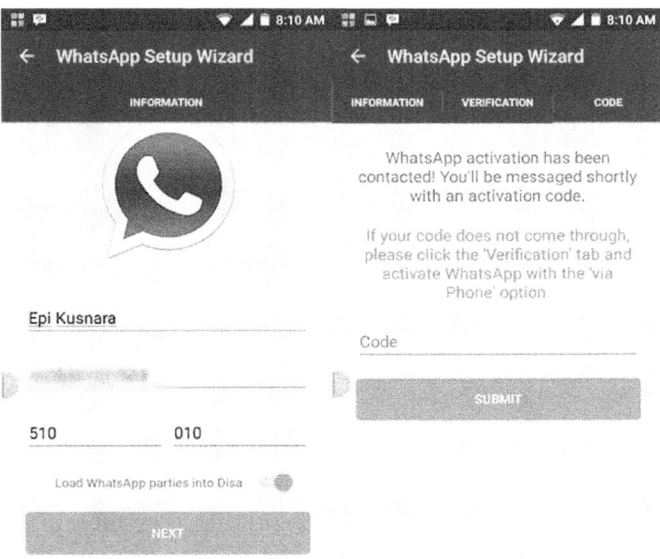

- Done. You can start using 2 WhatsApp numbers in 1 device.

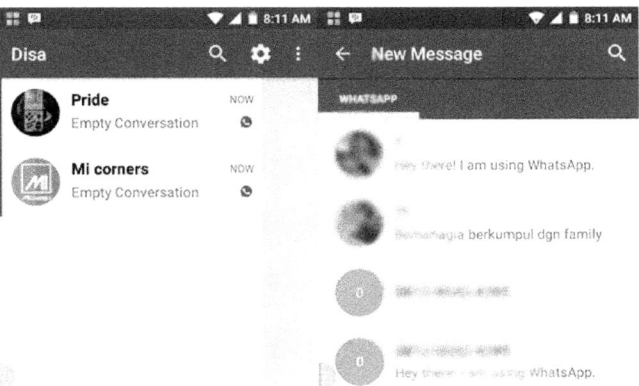

- You can perform advanced settings on your WhatsApp account by selecting **the Settings** button, and then select the three points next to WhatsApp *plugin* to make any other changes you want.

Mastering 37 WhatsApp Tricks

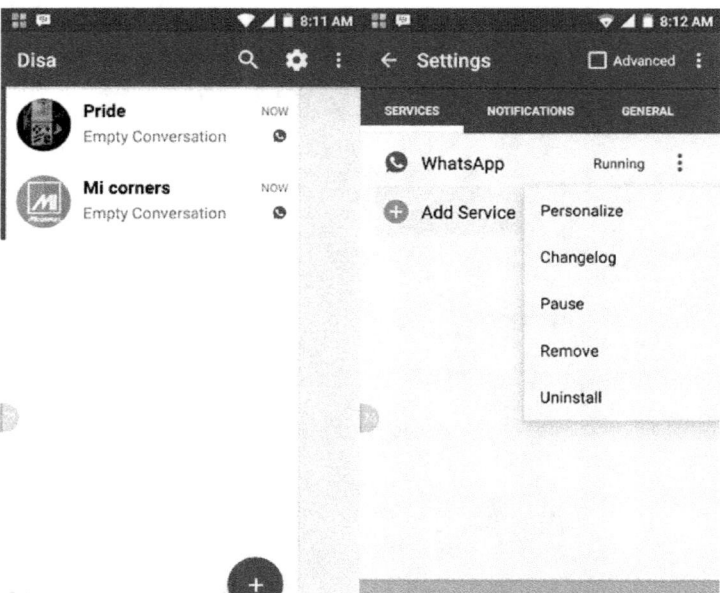

- Alternatively, you can click on the **WhatsApp Plugin** name if you want to arrange your **WhatsApp Profile.** You can also block numbers in Disa.

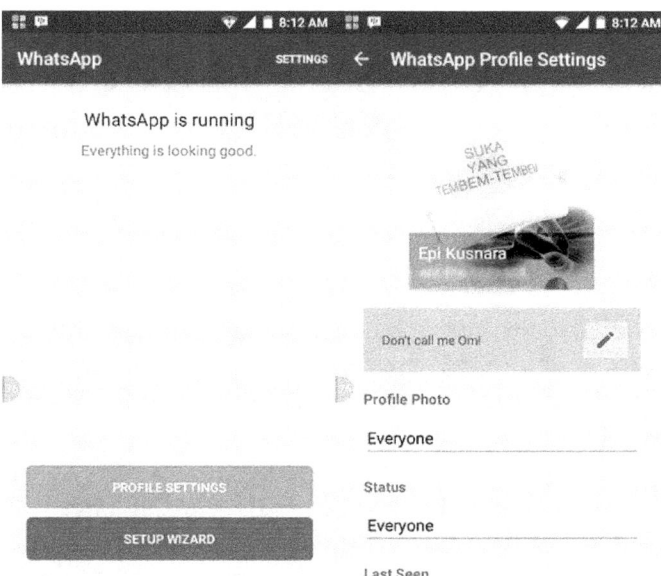

How easy to maintain 2 WhatsApp in 1 device using this Disa? No need to fear the license runs out as like using OGWhatsApp or other applications that always failed to install. More simple, and it is officially available on the Google Play Store. Good luck!

27. TIE WHATSAPP TO YOUR MOBILE NUMBER

Especially for some businessmen, business travel is often the case. And a new SIM card may be required for communications. Some people may faff about adding contacts one by one. In fact, you can make it simple.

When you changed to the new SIM card and activate WhatsApp, it may ask for registration with the new number. You can ignore this popup and use WhatsApp with the old number while your phone is inserted with the new number.

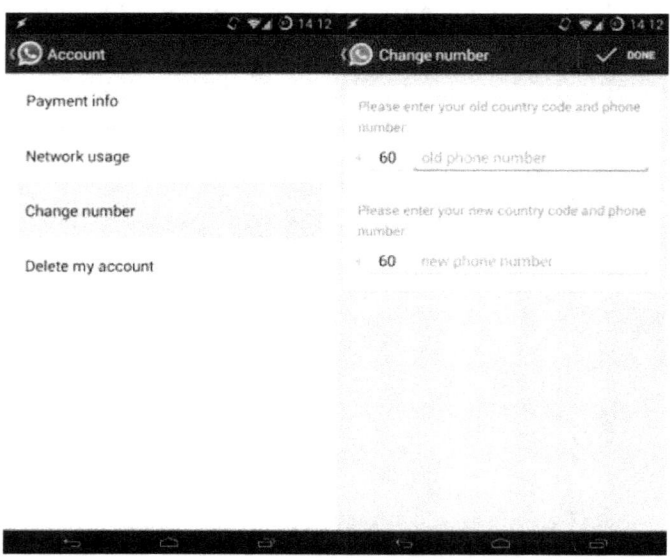

28. MAKE YOUR WHATSAPP FRIEND DAMAGE (CRASH) AND UNUSED

One *bug* found in the latest WhatsApp, it is not able to accommodate too many emoji. You should send approximately **4200 emoji** simultaneously to make your friends WhatsApp *crashes* and can not be used temporarily.

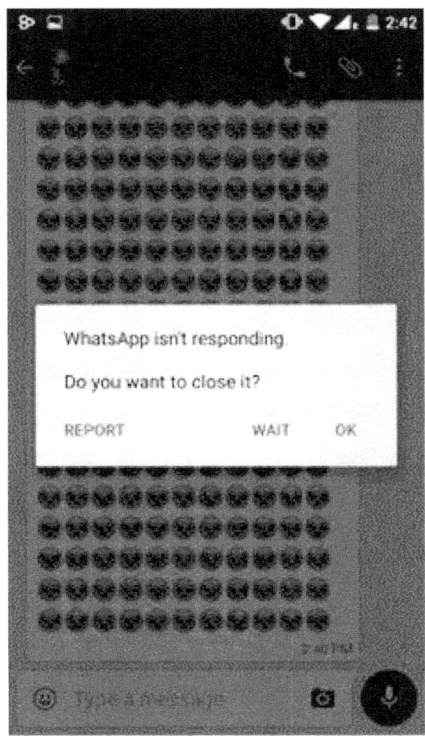

29. PROTECT YOUR PRIVACY BY DISABLING PREVIEW

When we are commuting, our private messages are easily peeped by people around us. In case of privacy leaks, WhatsApp has added the feature of disabling Preview so that your message details will not be disclosed.

Go to Settings, Notifications and disable Show Preview. Then no one can accidentally read your private messages by glancing on commute.

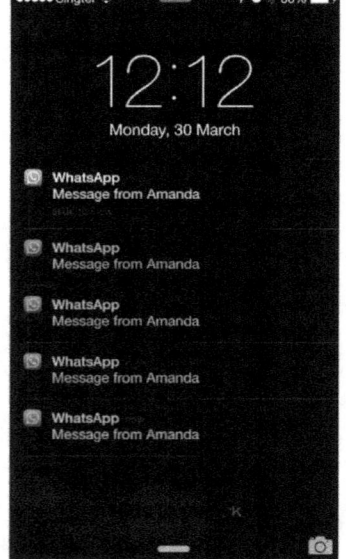 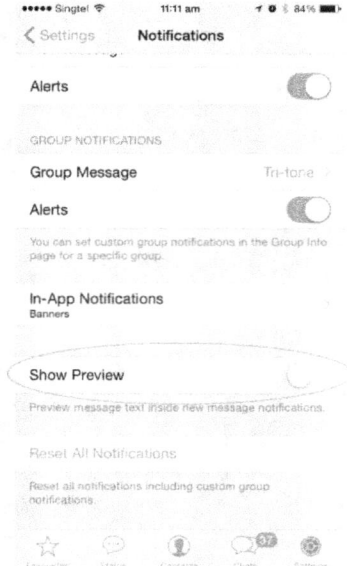

30. USE WHATSAPP ON COMPUTER OR LAPTOP

Not only can *chat* in WhatsApp via a smartphone, you can also chat using a PC or laptop. It's easy, you simply use the Web WhatsApp.

How do I synchronize with your Android WhatsApp WhatsApp Web?

- After you download and install the latest version of **WhatsApp for Android**, then through the menu on your new WhatsApp will show a new WhatsApp Web button. Press the button.

Mastering 37 WhatsApp Tricks

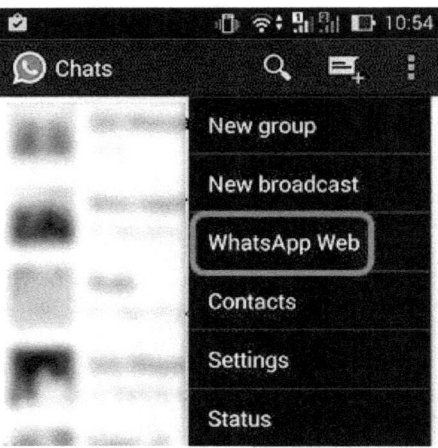

- Your camera will be active and in the meantime through the PC browser, open https://web.whatsapp.com/. Scan QR-code in the link using the Android camera.

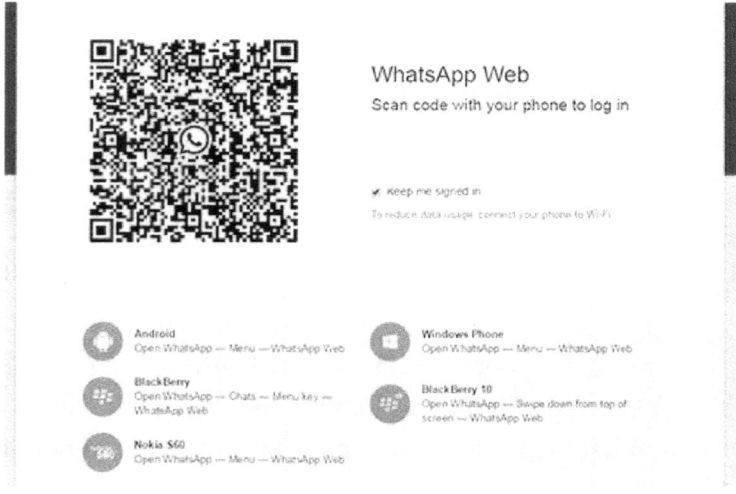

84

- WhatsApp in Android will automatically detect the QR Code and voila! Finally, you can chat through the browser on your PC. Good luck!

Mastering 37 WhatsApp Tricks

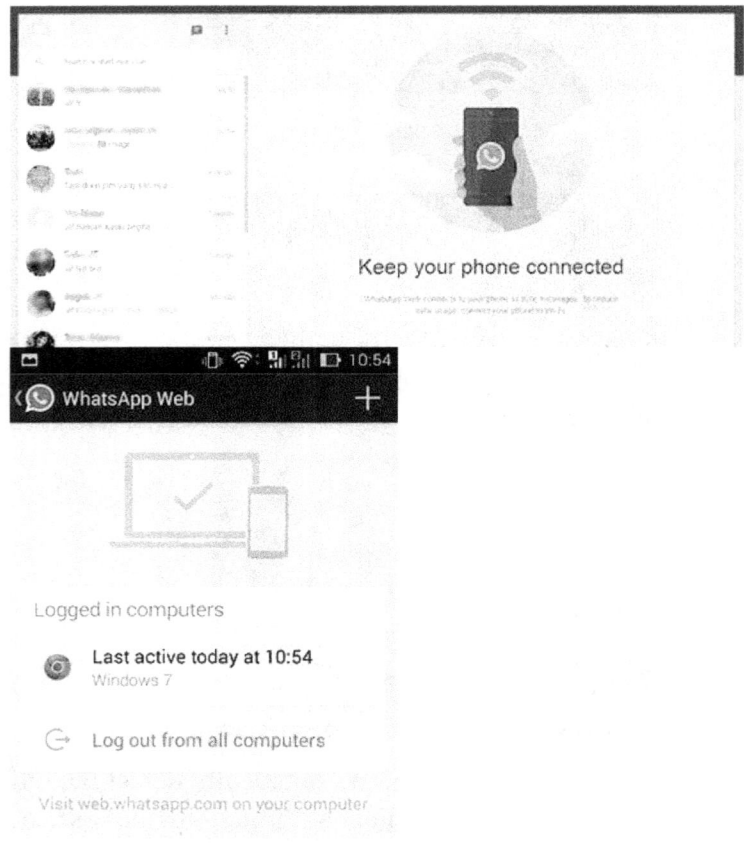

31. SENDING & SCHEDULING WHATSAPP MESSAGES AUTOMATICALLY

It turns out that you can send WhatsApp messages automatically and scheduled. It is fun, how to do? You need to install a third-party application called **WhatsApp Scheduler**. With this application, you can create a WhatsApp message, then set the delivery time and the number of times the message will be sent automatically.

Sometime you should wish a friend for his birthday or anniversary but are too tired to stay awake till the clock strikes exactly 12 am. Or like me, you always forget to wish the friend. Here's how to solve. You can now schedule WhatsApp messages on your Android phone using one of these apps which you will learn in this simple tutorial. This is a part of the WhatsApp tricks collection we shared some days back.

Scheduler for WhatsApp

1. Download Scheduler for WhatsApp from the Play Store.

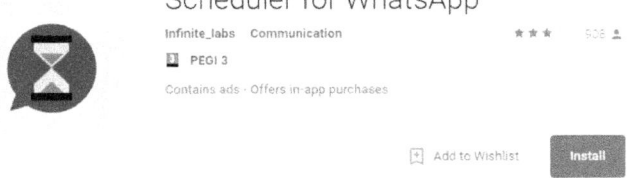

2. No need to root the phone. Works for non-Rooted phones also.

3. You need to enable accessibility service to use this app. Go to Settings->Accessibility and turn on the WhatsApp Scheduler.

Mastering 37 WhatsApp Tricks

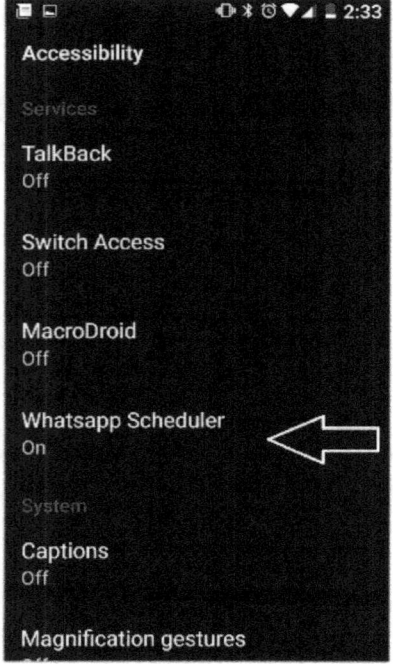

4. It has two icons at the bottom by which you can toggle between the pending and sent messages.

5. To schedule a new message goes to the pending messages and you will find a new message icon.

6. Enter the recipient's contact number. You will find all your WhatsApp contacts in the list and can add multiple contacts if needed. Group messages can also be scheduled using this app.

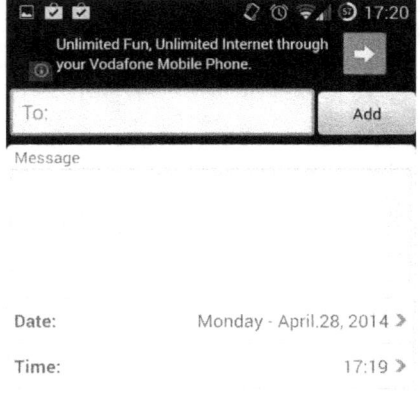

7. Write the message you wish to deliver.

8. Then fix the date and time for the scheduled message.

You have now scheduled the message. All your scheduled messages can be found in the pending messages.

Scheduled messages can be deleted here using the edit button. Once the message is delivered you will receive a delivery notification. This is a great app and does its intended tasks efficiently but the app is infested with ads which are very annoying.

Whatsapp Seebye Scheduler

1. Download WhatsApp Seebye Scheduler from google play.

Note: Just like the previous app this too works only on rooted android phones and it requires superuser access. After you grant access you can move on to scheduling messages.

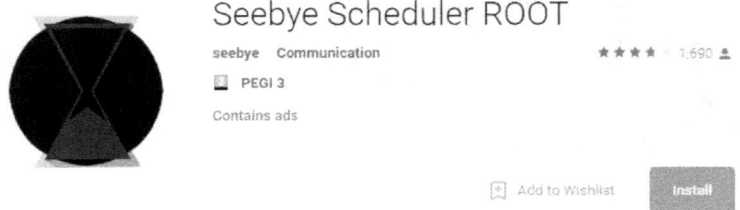

2. The user interface is very clean and simple. Just click on the '+' sign to add new messages.

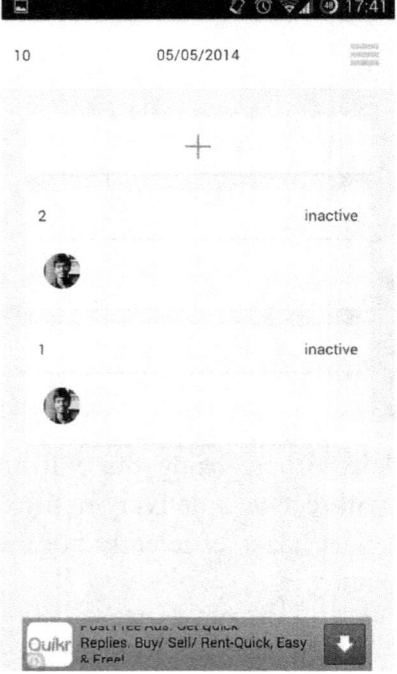

3. Next, you type the title of the message and message itself.

4. As in Message Schedule Lite, all the contacts are already synced and you just need to select them.

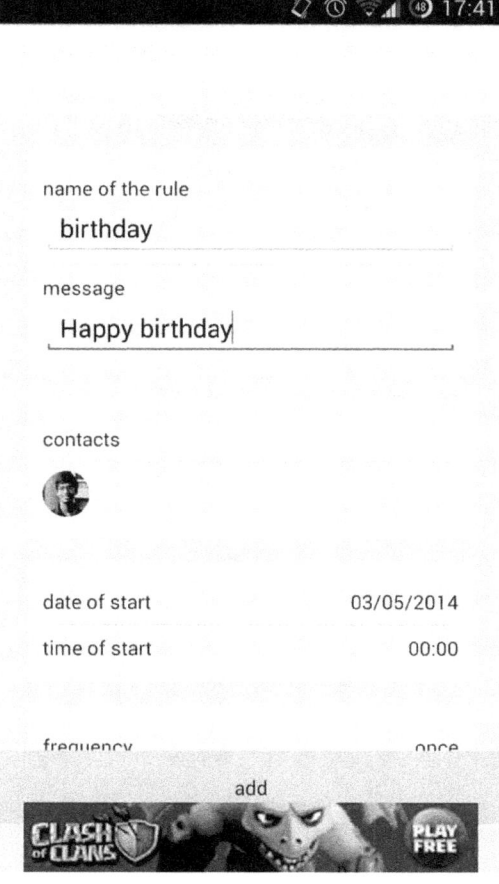

5. Set the date and time for your message.

Once that's done the app will return to the start screen where you will find all the messages scheduled. The messages that have sent successfully will have inactive written on them and active will be shown on the messages not yet delivered.

6. This app provides the added facility of setting frequency for your messages. So, if you want to send a message to a person daily or weekly, Seebye Scheduler will do it.

Even though both apps work quite similarly their small distinct features set them apart. The delivery alert of Message Scheduler Lite comes in handy as you wouldn't have to open the app to check the delivery status of your message. Whereas the ability to set frequency for the message will cater to the need of specific people. All in all, both apps work just fine and are ideal for scheduling messages.

32. HOW TO KNOW IF I'M BLOCKED ON WHATSAPP

You have already been explained on how to block WhatsApp contacts, then how do you know if your WhatsApp account is being blocked by others. It is not fully known. Through some WhatsApp feature, you can find out whether it is being blocked.

1. Check the Last Time Seen

Last Seen is a feature of WhatsApp to inform the user that their contact is online or to know when they were last online. Indeed, to see this last seen feature, each user must enable it. There are two possibilities if you are not able to see her last seen. Maybe she does not activate or indeed you are already blocked by her. If you able to find at her last seen, then you are not blocked by her.

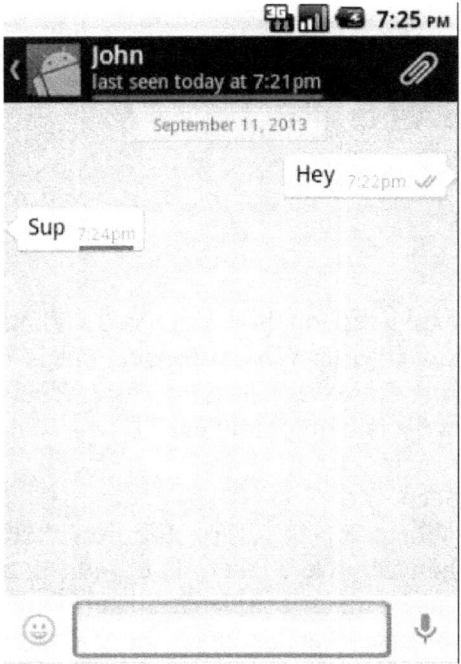

2. Check Blue Ticks

This blue tick could be phenomenal in WhatsApp because ultimately the user can know whether his chat has been read or not. Both sender and his contact must enable their read status on WhatsApp.

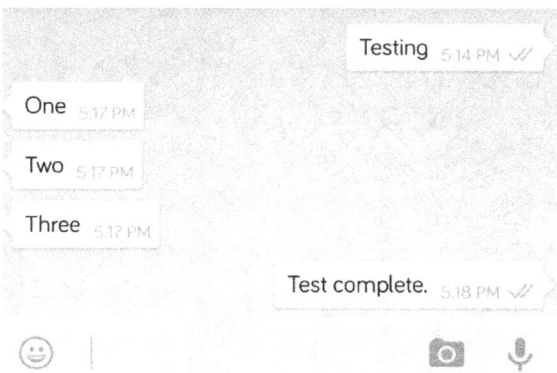

Of course, this has two possibilities as well, either he is not to activate or indeed you are already being blocked. If you send a message and the blue tick is checked, it means you are not blocked.

3. Phone Through WhatsApp

Well, maybe this way will a bit tricky because you must be ready for a topic of conversation when he picked up the phone. If you can not call him, it possibly has no signal, the number was changed, or your contact mobile phone was being damaged. But if these three marks happened, it seems likely you are already being blocked.

Try to meet him if you know his or her campus or his usual hang out place and ask directly.

33. MUTE GROUP CHATS IF YOU GET BORED

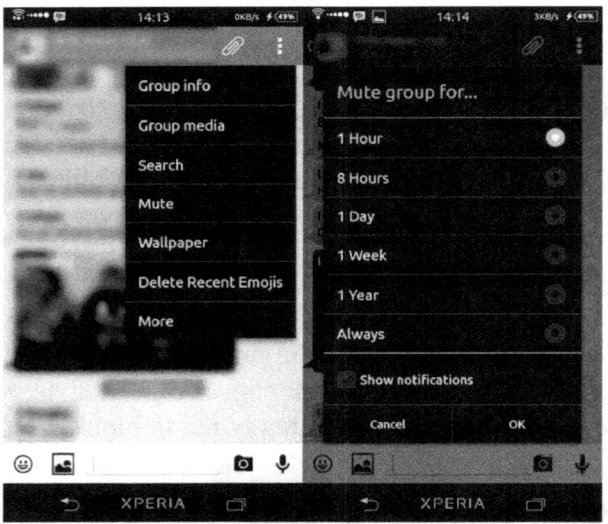

In the WhatsApp application, you probably have joined some group. Either voluntarily or out of necessity. Many things are discussed in the group. Starting from sharing vital information, send an invitation, up to unimportant things like joking, send MEME and funny video, or even spread gossip.

Notifications on WhatsApp group sometimes make some people uncomfortable. Although we do not currently want to chat within the group, still we will receive notifications. It is not a matter if the group notification is low. However, it will be a problem if you keep the smartphone to vibrate due to the notification, you will lose a little concentration because of such notification. If you feel being disturbed but unfortunately you do not want to leave the group, you can turn off these notifications.

To stop notifications of WhatsApp group, you can use the following ways:
- Open one WhatsApp group that you want to stop notifications
- Select the menu in the upper right corner, then select Mute
- You can set the time to mute the notification. Do not forget to uncheck the **Show notifications** so that the group notification is hidden.

To restore it, you just need to return to the menu and select unmute. Good luck!

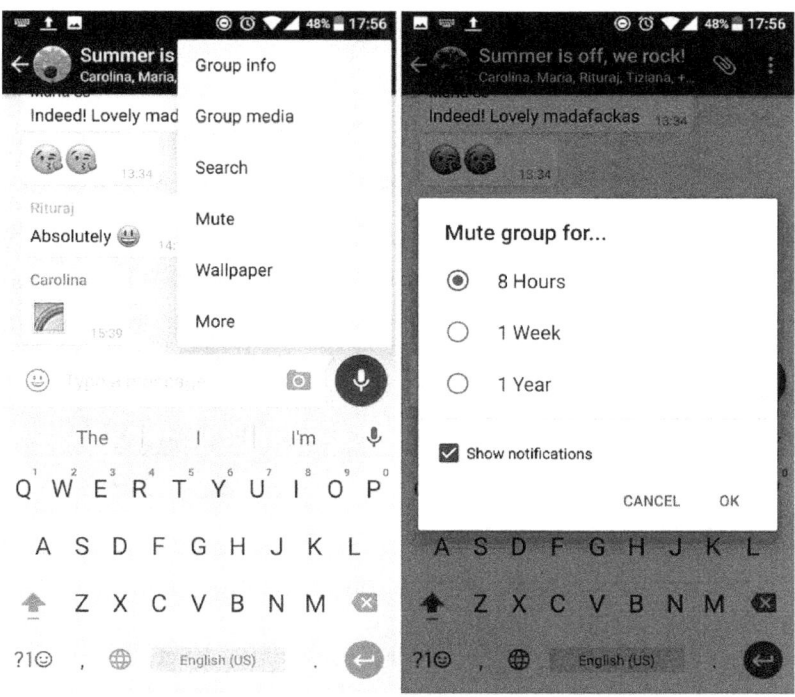

34. BACKUP WHATSAPP MESSAGE TO CLOUD STORAGE

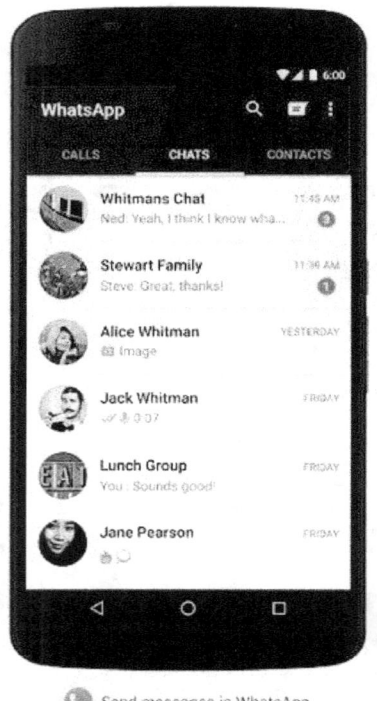 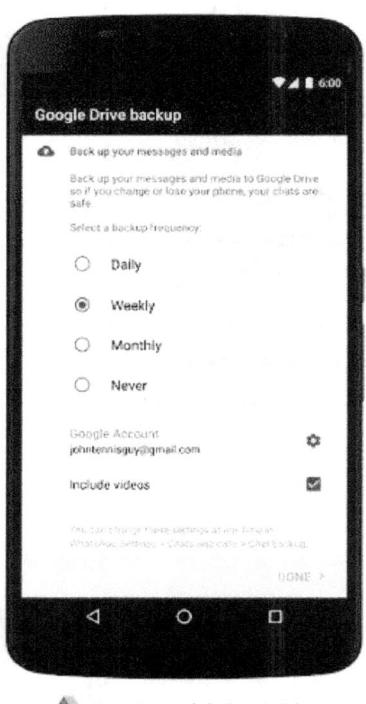

Send messages in WhatsApp — Keep them safe in Google Drive

You can use the Sync application to backup all the data on your WhatsApp to various Cloud Storage. This is very useful if you want to change to a new smartphone. Previously, we have discussed ways to backup WhatsApp data to Google Drive. Apparently, in addition to Google Drive, you can backup the data WhatsApp to other cloud services such as Dropbox, OneDrive, Box and another cloud storage media.

To backup WhatsApp data to anything other than Google Drive, we can use Sync application by Resilio:

Download Resilio Sync and install as usual on Android.

Open the application and enter your identity to be easily recognizable.

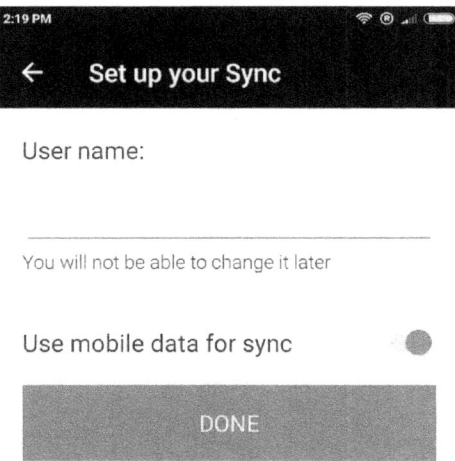

Pick the folders you want to sync across computers. Click the "+" button and select Add backup.

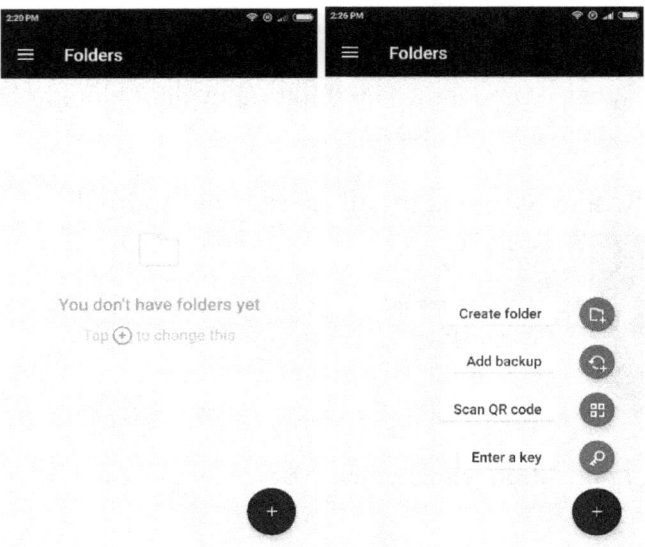

Find WhatsApp folder and select Choose Folder, then tap Add.

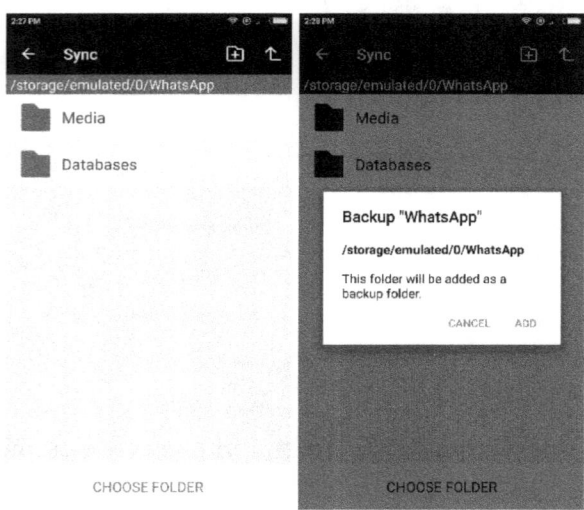

After selecting the folder, select the Share menu> Mail (or any available option you like) to get the link that will be used in the next tutorial.

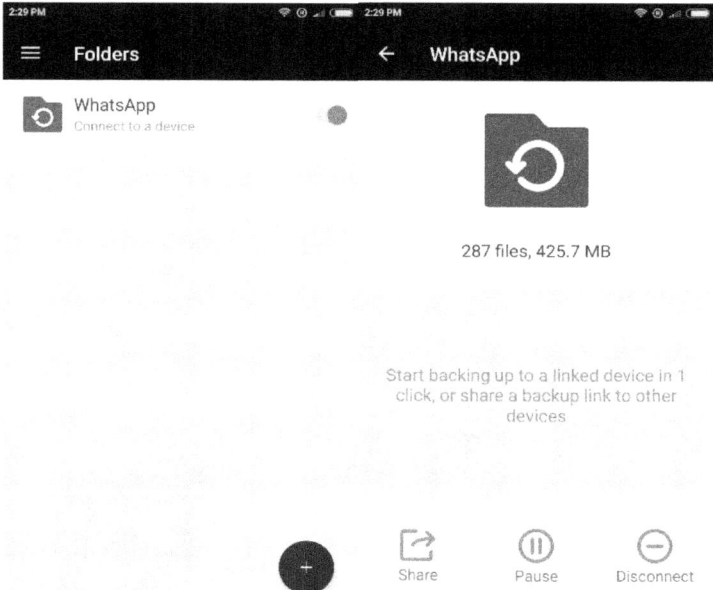

Download Resilio Sync software from https://www.resilio.com/individuals/. Locate the free download menu then select free download for home use. Install the software on your computer or laptop.

Mastering 37 WhatsApp Tricks

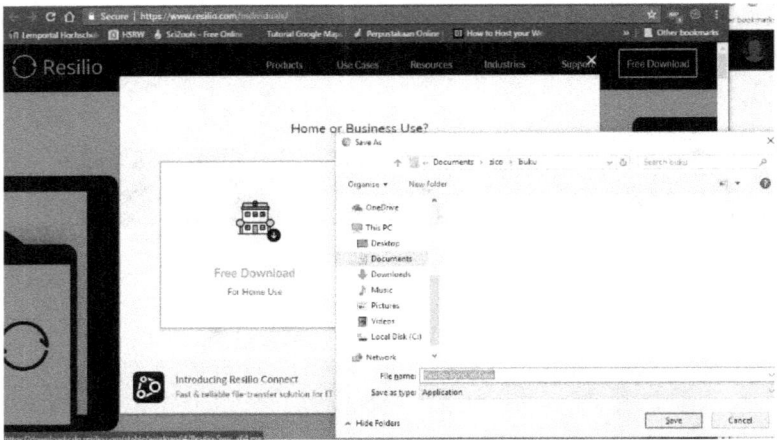

Locate gear icon menu and select Manual connection ....

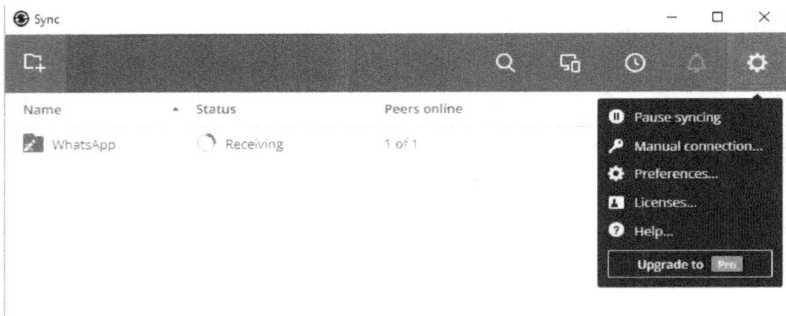

Enter the link that is already being copied from email, click the Next button.

Open the Sync application on Android and locate the Notification icon, then click the check button.

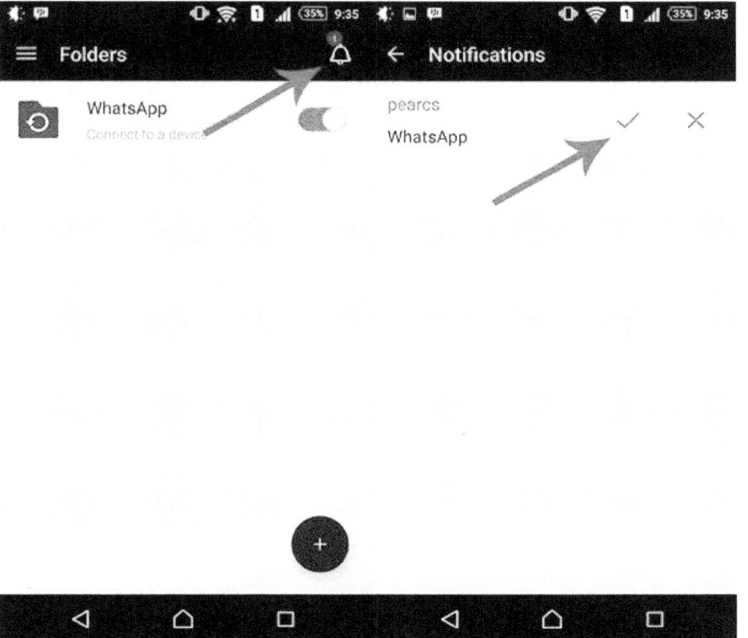

WhatsApp folder will automatically be synchronized with your last Sync software on the computer.

For uploading to Dropbox, Box, Copy, or another cloud storage media. You can copy the entire WhatsApp folder in your Sync computer.

35. USE WHATSAPP WITHOUT MOBILE NUMBER

Apparently, you can *chat* with WhatsApp with devices that do not have a SIM card, such as a tablet. Open Google Chrome on an Android device, then select Settings and then tick the Request Desktop Site.

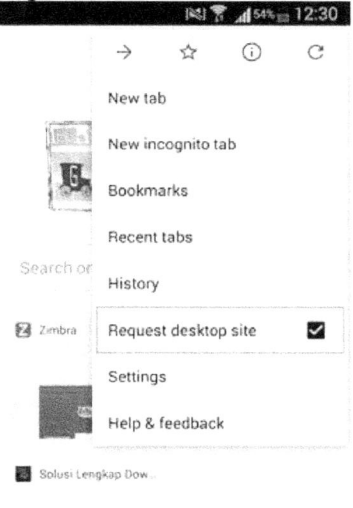

Access to http://web.whatsapp.com

On the other device that already has a WhatsApp account, open the WhatsApp Messenger and select Web, and then scan the QR Code on the previous device.

After scanning the QR code, and you can use WhatsApp Messenger on devices that do not have a SIM card, such as tablet that only has Wi-Fi connection.

The concept is like how to use WhatsApp on PC, but use a mobile device.

36. MAXIMIZE WHATSAPP NEW FEATURES

Many cool features are presented in WhatsApp latest *update*. As *a preview of* website *links, custom notifications,* and so on.
- Clear Chat Easily

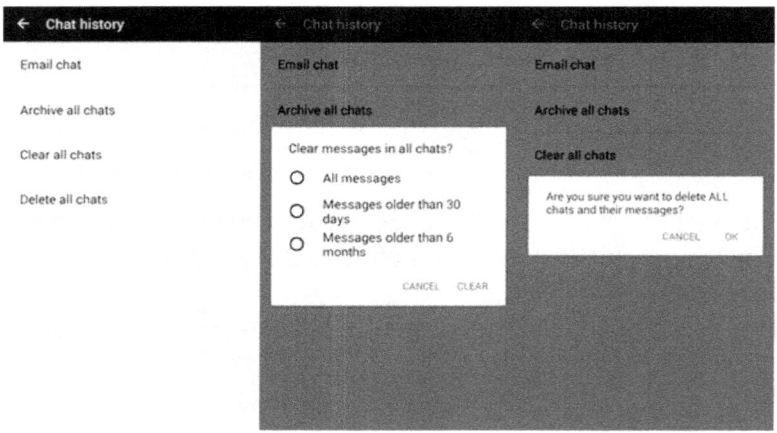

If you do not want to save any old or sensitive messages, WhatsApp has new features that can make it easier to remove the message. Open Menu> Settings> Chats and Calls> Chat History> Clear All Chats. There will be three options, i.e. delete all messages, messages older than 30 days, or messages that are older than 6 months.
- Mark Messages Being Read or Unread

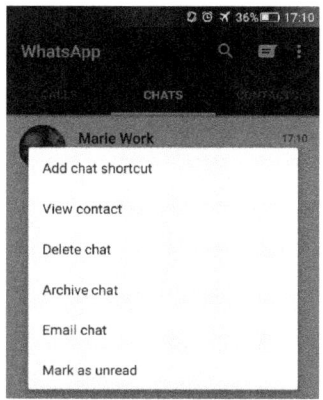

If you accidentally open a conversation, but do not have time to read it, then you can restore status to Unread by long pressing the conversation, then select Mark as unread. You can also do the opposite with unread messages but want to be marked as read in the same way.

Use Custom Notifications for People or Group

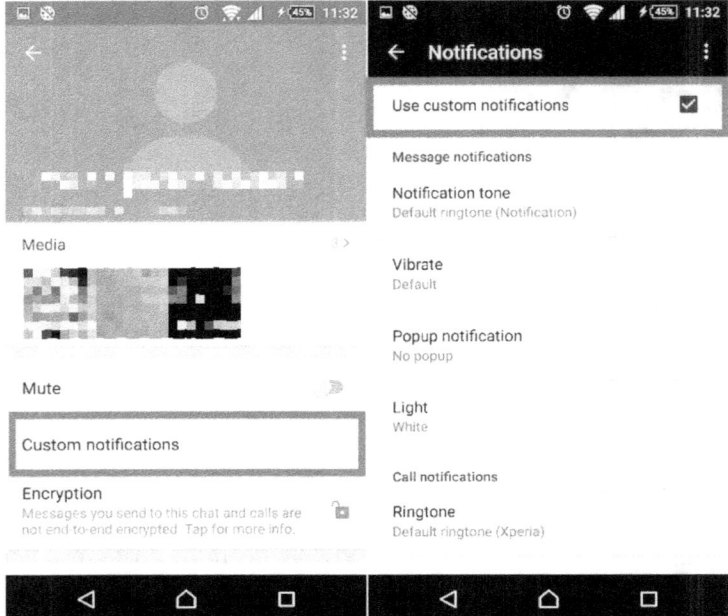

Some people or groups indeed are more important than others. Therefore, WhatsApp present custom notifications feature so that specific people or groups that you want, have a different notification to another.

Preview Link

This feature is also interesting to observe. Now if you share website links in WhatsApp, then it will show a thumbnail preview below the title of the link. So, you can see at a glance what it is contained in the link that somebody gave.

37. FIVE FRAUD TYPE STRIKE WHATSAPP USERS

As the most preferable *chat* application, of course, there are bad people that utilize the WhatsApp names of to deceive others and attempt to reap a profit from them. Hence, through the following article, we would like to inform you. Go ahead and read the following tips on 5 Dangerous Fraud Type in the name of WhatsApp.

1. Fake Voicemail

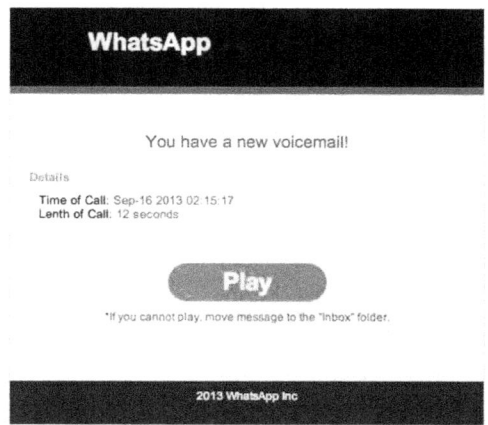

Fake voicemail is one of the traps created by *hackers* who want to take valuable information from your smartphone. They sent an email message that there is a voice message (voicemail) via WhatsApp that you have not heard. To listen to it, you are asked to press the **Play** button in the email. However, the key was apparently connected to a *malware* that can steal personal information and even lock the smartphone you are not responsible.

2. WhatsApp Gold Edition

Do you want to look attractive and elegant? Circulating on the internet WhatsApp MOD version called **WhatsApp Gold Edition.** This WhatsApp indeed looks golden, with additions of new features such as emoji and special background. However, it requires you to pay **40 dollars** every month to continue using WhatsApp Gold Edition. Useless once.

3. WhatsApp's Spy

At Google, there are many websites that offer services to hack an account or WhatsApp *chat* of others. But the thing to remember is, among all these services, most of which are malware that is ready to steal personal data from your favourite smartphone. So initially, you want to know others personal information using this hacking applications. In fact, your *smartphone* becomes their victim.

4. $ 500 Starbucks Gift Card

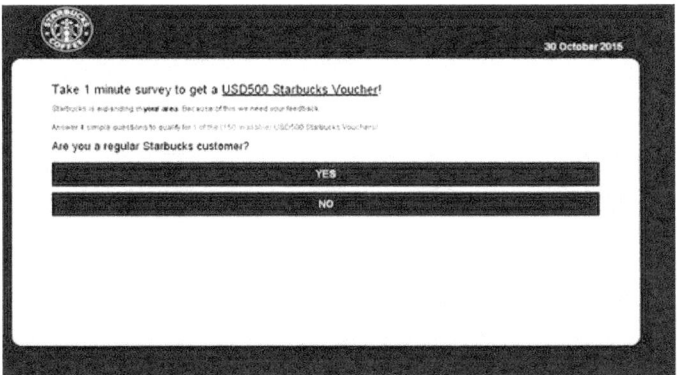

There is a scam that is circulating on the Internet on behalf of **Starbucks,** and it asks its readers to participate in the survey. The fake survey derived from the irresponsible party is trying to steal personal data, and growing each day with several variants. There on behalf of **McDonald's, IKEA, H&M, KFC, 7-Eleven,** even **Zara.** Beware!

5. Investment Fraud

Some time ago it had circulated a form of fraud that attacking WhatsApp users. The scams contain a call to invest into a company called **Avra Inc.** The message claims that every investor who put the money in Avra will benefit up multiple times. Unfortunately, the temptation was merely a trick by certain parties. *Scammers* who are responsible for these scams will deceive every investor and suck their money by playing the stock value.

So, there are 5 frauds -claim on behalf of WhatsApp- that you need to know. We hope that you are using WhatsApp to take care of various scams circulating.

***

ABOUT THE AUTHOR

Zico Pratama Putra, M.Sc. is an active engineering practitioner and an computer science book writer. After graduating from Institute of Technology Bandung (ITB) in Aerospace Engineering, he spent years working as an IT consultant and telecommunications. In 2011, he continued his Master studies to Germany in the Rhein-Waal University of Applied Sciences while working in the field of computers in one company in Germany. He continued his PhD at the Queen Mary University of London in Human Computer Interaction field. He writes and speaks at various events in Germany & UK.

Find out more at https://www.amazon.com/Zico-Pratama-Putra/e/B06XDRTM1G/

CAN I ASK A FAVOUR?

If you enjoyed this book, found it useful or otherwise then I'd really appreciate it if you would post a short review on Amazon. I do read all the reviews personally so that I can continually write what people are wanting.

If you'd like to leave a review, then please visit the link below:
https://www.amazon.com/dp/B06XS99PKP
Thanks for your support!

INDEX

A

accessibility, 8, 87
automatic, 42

B

Backup, vii, viii, 10, 20, 24, 52, 53, 100
Block, 43
blue tick, 31, 33, 34, 35, 62, 63, 64, 66, 96, 97

C

command prompt, 57, 59

D

Drive, 52, 53, 55, 100, 101

E

Excel, vii, 46

F

Fake, 114, 115

I

Instagram, 4

IP Address, 57

L

Last Seen, vii, 32, 33, 95

P

Phone Numbers, 16
PowerPoint, vii, 46
Privacy, 31, 32, 64, 66

R

Restore, 13, 54
roaming, 28

S

safety, 5
Snapchat, vii, 4, 36, 37
star, 38
    Starred Messages, 38

U

unread, 111

V

Video, vii, 44

www.ingramcontent.com/pod-product-compliance
Lightning Source LLC
Chambersburg PA
CBHW071447180526
45170CB00001B/498